Published by Lexington Books
An imprint of The Rowman & Littlefield Publishing Group, Inc.
4501 Forbes Boulevard, Suite 200, Lanham, Maryland 20706
www.rowman.com

Unit A, Whitacre Mews, 26-34 Stannary Street, London SE11 4AB

British Library Cataloguing in Publication Information Available

Library of Congress Cataloging-in-Publication Data

ISBN 978-0-7391-9668-7 (cloth : alk. paper)
ISBN 978-0-7391-9669-4 (electronic)

∞ ™ The paper used in this publication meets the minimum requirements of American National Standard for Information Sciences Permanence of Paper for Printed Library Materials, ANSI/NISO Z39.48-1992.

Printed in the United States of America

Dedication

This work is dedicated to the kind people of Tono, their mayor Toshiaki Honda, and to all leaders, staff, and volunteers at Tono Magokoro Network. They continue to be a lifeline for victims and countless survivors of Japan's March 11, 2011, disaster. As in the meaning of *magokoro*, they embody "true heart."

Table of Contents

Acknowledgments

Writing this book on Japan's March 11, 2011, earthquake-tsunami-nuclear meltdown has been for me a private struggle, yet one that only comes from shared labor. Without others who matter, life is generic. Writing is as well. Alex Haley wrote that "Every death is like the burning of a library." Many of us may disagree. Would not my death be the end of a book, not library? Yet, each book rests upon the shoulders of other books. Our journeys interpenetrate and run deep. And with each fresh day, new discoveries—delicate, difficult, delightful, disturbing, and in-between—shape for us a renewed soul.

I put together these initially scattered thoughts after two sustained visits to my homeland, Japan, after its tragic triple disaster. My first visit was in 2012. With generous support from the East Japan Railway Culture Foundation and assistance from its manager Kunio Aoki, his assistant Mizuki Naito, now Abe, and Hiroko Suzuki, who then managed Tono Furosato Village, I spent the early part of winter in Tohoku, the region directly affected by the disaster, where I met and spoke with survivors, victim's families, and key persons involved in recovery. There I first met Bunzo Kanayama, Jin Imai, and Akiko Iwasaki, whom I am now honored to call my friends.

For my second visit, again in the dead of winter, this time in 2014, I wanted to get more deeply into the skin of those who faced 3.11's horrific destruction—survivors eking out each day at a time, along with fully involved city leaders and those actively engaged in recovery work—by spending more time in the still afflicted areas, particularly along the Sanriku coastline. I especially needed to be there during the March 11 memorial services. At first, I was not sure about volunteering. I simply wanted to *do something*, work in whatever way to more intimately understand others' plights. Hiroko, Kunio, and Mizuki strongly encouraged me to volunteer. They supported me

all throughout, enabling me to have some of my most memorable lived experiences, a few of which I am privileged to share with readers. This experience "on the ground" will always burn within me. For Kunio, along with his wife Izumi and precious family, Mizuki, and Hiroko's steadfast help, guidance, and encouragement, I remain forever grateful.

Again, it was through the invaluable assistance of East Japan Railway Culture Foundation that I was able to travel throughout this part of Japan and experience the eye-opening encounters that have helped me to truly see, acknowledge, and learn. My teachers include many survivors, volunteers, and staff, as well as leaders like Tono City Mayor Toshiaki Honda and Minamisoma City Mayor Katsunobu Sakurai. I've also learned from respected scholars and experts along the way including Makiko Okuyama, Director of Psychosocial Medicine at Tokyo's National Center for Child Health and Development, Naoko Kakee, Chief of the Division of Health Policy and Bioethics at the same National Center, East Japan Railway's Toshimi Kushida and its Culture Foundation's Noriko Yamamoto and Ichiro, Katsunori Kawano and Naoki Hayashi of Fuji Xerox, Professor of Occupational Therapy Sayoko Kawabata of Bunkyo Gakuin University, and Professor of Philosophy Tomoaki Tsuchida at Waseda University.

I give thanks to all my many Japanese friends and colleagues who continue to encourage me. They include Noriko Nagao, who teaches at Kobe University's Graduate School of Health Sciences, Carl Becker, Kyoto University Professor of Ethics and Comparative Religions, Yoshinori Hayashi, Satoshi Kodama, and Misao Fujita, all teaching and doing research in Kyoto, and Keiichiro Yamamoto, who teaches at the Center of Biomedical Ethics and Law, University of Tokyo. As always, I give special thanks to my dear friend Akira Akabayashi, who founded and directs the University of Tokyo's prestigious Center of Biomedical Ethics and Law, the first center of its kind in Japan.

My 2014 stint in Japan was also possible thanks to The College of Saint Rose graciously granting me a sabbatical for that semester to conduct further research. My heartfelt thanks to Interim President Maggie Kirwin, Provost Hadi Salavitabar, and former and current Deans Kim Middleton and Leroy Bynum for their enthusiastic and reliable support in enabling me this opportunity to better understand and appreciate the complexities of the tragedy and aftermath. I am further grateful to the college's new president Carolyn Stefanco for her marvelous presence, leadership, and faith in my work, as well as to Father Chris Degiovine, former Dean of Spiritual Life, and my department, especially philosophy and ethics faculty Laura Weed, Jeff Marlett, Bruce Johnston, and Jean Wiley for their authenticity and support. My kind gratitude also goes to administrators at St. Bernard's fine School of Theology and Ministry in Albany: Dean Sr. Kitty Hanley, Sr. Damien Cushing, and Eileen Davenport. And to my special mentor here in Albany, the ever-sharp

and kindhearted Sr. Rose Regina Smith. For me, no gratitude is complete without expressing my heartfelt thanks to R. Mark Sullivan, former president of The College of Saint Rose, and his precious wife Kathy. They both remain unfailing friends and pillars of support to my wife Brooke and me.

As part of my research, I traveled to Norway to learn more from key persons involved in the response to and recovery from the brutal July 22, 2011 Oslo explosion and ensuing carnage on the nearby island of Utøya by a lone gunman. They were invaluable in helping me to better understand the influence of Norwegian culture and worldviews in its disaster response, even though Norway's disaster was directly human-incurred, in contrast to Japan's 3.11 blend of natural and technological calamities. I remain especially grateful to Trondheim University psychiatrist Are Holen, who played a key role in the trial testimony. I also thank my faithful friend and colleague Professor Julie Feilberg, who teaches applied linguistics at Norwegian University of Science and Technology, for introducing me to Are. My deep gratitude also goes to Oslo Red Cross Director Anne Cecilie Fossum, who orchestrated aid and response from support groups, and Alexandra Bech Gjørv, who led the Independent Commission charged with investigating the tragedy. Other wonderful voices come from Trondheim University linguist scholars and professors Gøril Thomassen and Hanna Rustad. Ivar Vaksdal and I met at a café in Burlington, Vermont, where he shared with me insights from his small town in Dalekvam.

To be a volunteer following disaster is to take a detour from the everyday. I needed a shift from academia's routine, one that would enable me to take a less traveled yet more direct route to better grasp others' suffering. Along these lines, I offer my heartfelt gratitude to volunteers and staff with whom I worked at Tono Magokoro Network. My thanks extend to Tono Magokoro Net's leaders, Chairman Kazuhiko Tada, and Vice-Chairmen Ryoichi Usuzawa, Yusuke Kotani, and Keiko Maekawa. My kind thanks especially to Magokoro Net's Public Affairs Manager, Makoto Yanagisawa, for furnishing details of the organization's history.

None of our efforts as volunteers would have been possible without the dedicated work of staff members who work behind the scenes, often daily in offices right next to our meeting room. I would venture to their offices when registering for activities or reserving a washer and dryer for my clothes. The Net's invaluable staff includes Directors Koji Kise, Masahiro Sasaki, Hideki Matsunaga, Toshihiro Kuramoto, and Yoshiya Anbe, its Auditors Ryoji Arata and Hiroshi Sasaki, its Counsels Eietsu Arakawa and Fumio Terui, and Managers Kanako Hosokawa, Chikara Oyama, Fumiko Yanagi, Yukie Sasaki, Chihiro Yamamoto, and Chie Kotani.

My special gratitude to the many volunteers I've had the pleasure to work with. I still stay in touch with my good friends like Kyoko Akasaka and Ayaka Nishii, both from my birthplace Fukuoka, Masashi Tanoue, Michiyo

Chiba, Nobu Kobayashi, Katsunori Kumagaya, Dr. Shinichiro Fujikawa, Uenishi Yukata, Komukai Hiroto, Toshimi Kushida, Takashi Ichioka, Tomomi Kaneko, Yumiko Murakami, Yoko Kanda, and Yosaku Oshiro.

My winter 2014 visit was the longest sustained period of cold I've encountered. After all, it was Tono's brutally cold season, and, after my triple hernia repair in December I needed to be extra careful. Throughout this enduring chill my good friends' concern for my health showed in their displays of support, like Hiroko lending me her traditional Japanese hot water heater, and Mizuki's gift package from Tokyo of hand and feet stick-on warmers. Hiroko would occasionally meet with me after our late afternoon group meeting for dinner. She would then help me pick up my next days' food supplies at the local convenience store, expertly maneuvering her car through Tono streets' snow and ice. Hiroko often introduced me to her friends, like Ichiro Kobayashi of Tohno Seimitsu, as well as an unforgettable local performance of Miyazawa Kenji's "Restaurant of Many Menus," in which the acting was superb. Hiroko's generosity knows no end. Even now, she is kindly housing my sleeping bag and boots for my next visit.

Throughout the course of my research, numerous friends and colleagues have encouraged me in so many ways. I've often been invited to speak about my experience. Joe Nash at our local Colonie Library always extends an invitation for me to share with the community. When honored to deliver the Dr. William T. Sirridge Medical Humanities Lecture at the University of Missouri-Kansas City School of Medicine, I spoke on Watsuji Tetsuro's notion of "in-betweenness," or *aidagara*, in the context of the Japanese disaster. Dr. Marjorie Sirridge inspired me through her singular scholarship and presence. I can never forget my beloved friend and mentor Marjorie and our conversations about literature, especially Tolstoy, and medicine. She passed away in 2014, and my work on the importance of presence in healthcare is always written in her memory. Marjorie's sons Dr. Christopher Sirridge, esteemed specialist in hematology and oncology, Pat and Stephen, well-respected lawyer and psychologist respectively in Kansas City, and daughter Mary, Professor of Philosophy at Louisiana State University, all showed their strong support. I will also always cherish my dear friend Marilyn Pesto, Director of the Sirridge Office of Medical Humanities and Bioethics. Marilyn is never without a smile, one of the kindest people I know, and with saintly patience. She never ceases to bless and reassure through her steady guidance, counsel, and friendship.

It was for me a marvelous sharing experience when I was honored to deliver the annual Semrad Lecture on Non-Western Thought at Creighton University. I spoke about my research, and colleagues, now friends, in the Department of Philosophy and the East-West Studies Program favored me with their curiosity and penetrating questions about Japanese culture and its impact on the March 11 tragedy. I deeply thank Amy Wendling, Philosophy

Department Chair; Professor Richard White, who embodies Zen centeredness; Dr. Maorong Jiang, Director, Asian World Center; and Dr. Jinmei Yuan, scholar at the center. My conversations with Dan Semrad sparkled with his genuine interest in intercultural studies and particularly matters related to Japan. I thank him for his enthusiasm, attentiveness, and empathy.

The annual international Communication, Medicine, and Ethics (COMET) conference was another major venue where I shared my research. My friends and colleagues who reunite there remain a steadfast source of encouragement. For their top-notch scholarship and sterling example, I am always especially indebted to Srikant Sarangi, Director of Aalborg University's new Danish Institute of Humanities and Medicine; Peter Schulz, Director of the Institute of Communication and Health at the University of Lugano; and Ulla O'Connor, Director, Indiana University Center for Intercultural Communication.

Many others include José Carlos Gonçalves at Brazil's Universidade Federale Fluminense for his marvelous work on the meaning of presence in healthcare; Myra van Zweiten and her inspiring work in healthcare ethics at Amsterdam's The Good Worker; Jill Dales, Director of Communication Skills at Newcastle University; Kim Grego, professor of linguistics with whom I had the pleasure to work during her Fulbright Research stay at The College of Saint Rose. Kim conducted research on elderly perspectives regarding online healthcare information and kindly stepped in for me when I was recently ill. And there is Alessandra Vicentini, another linguist scholar, and fabulous singer, at the University of Insubria who also visited Saint Rose for research; Roxana Delbene who teaches at New York City College; gracious Ulla Hellstrom-Muhli, a communications and health professor at the University of Skvode in Sweden; Eleni Siouta, Health and Welfare doctoral candidate at Jönköping University in Sweden; Anne Storey, linguist at the University of Hong Kong; and Iréne Josephson, Research and Development Office and physiotherapy scholar at Jönköping University. My chorus of gratitude extends especially to Marie Manidis, Postdoctoral Research Fellow at the University of Technology in Sydney, who shares a special interest in Japanese studies; Antoinette Fage-Butler at Aarhus University, Denmark, for her wonderful contributions in health communication; my wonderful and supportive friend Olga Zayts, linguistics scholar at Hong Kong University; and Mei-Hui Tsai, scholar in medical discourse at National Cheng Kung University, Taiwan, whose special focus is on the doctor–patient relationship.

I remain forever grateful to my dear friends and marvelous scholars, Renée Fox at the University of Pennsylvania, and Kay Toombs at Baylor University. They both embody for me what it means to be a wise scholar. The same goes for Paul and Eileen Churchill, faithful and truly kind-hearted friends and teachers. Former philosophy colleague and steady friend Dr.

Frank McCluskey and Melanie Winters, now newlyweds, have always been in my corner, sharing with me their invaluable ideas and input. The same goes for Bonnie Gray and Ari Greyson, marvelous professors at Scottsdale Community College. Ever since they've invited me there to share my research, we've kept our friendship strong. So also with always kind Diana and Eric Carlson in Arizona. Nobuko VanKameke in Maine continues to inspire through her elegant calligraphy and example. My lifelong friends—I can think of no better friends for life—Jim Fanning, Tad Connerton, Cliff Muldoon, and Ron Brooks have stayed solid and always there. Through all these years, their encouragement and support remain unbroken.

As for the staff of Lexington Books, I cannot say enough. As always, I remain grateful to my editor Jana Hodges-Kluck for her unwavering encouragement, patience, and understanding. I offer special thanks to her former assistant Kari Waters, wonderful and hard-working production editor Shelby Kennedy, and the splendid staff who care far beyond simply dotting 'i's and crossing 't's and continue to promote what's worth writing about. Deep thanks go to my graduate assistant Katherine Tarr who offered help overseeing ethics sessions here on the campus while I was away in Japan. Gratitude also goes to Jay Jochnowitz, Opinion Editor for the *Albany Times Union*. He has always graciously supported my work and more recently through granting a temporary reprieve for me to finish my manuscript. Strong measures of thanksgiving also to our dear friends David and Deb Phaff, who are always there for us to lend their kind help. I also express my heartfelt thanks to Bill Borman, who graciously and most arfully designed the front cover image in his woodblock style. A superb artist and good friend, may his exceptional gift bless his journey.

We measure any degree of fracture from its earlier state. Despite Japan's past decades of unsettled employment and dissolving personal, social, and national security, it has managed to maintain national stability. Yet 3.11 brought with it a cataclysmic rupture from the somewhat stable to the unhinged.

The 3.11 earthquake, tsunami, and radiation fallout pounded upon the Japanese psyche what its people have always known. And what is true for all of us. We all live on the edge. In the past six years, Brooke and I experienced our string of personal earthquakes when all four of our parents died, one after the other. Immediately after 3.11, I made a silent pledge to my departed mother, Misae Kimura, to learn more about the tragedy and share what I learn with others. This offers me a necessary pause, my own crack in my everyday discourse of healthcare ethics, intercultural studies, and philosophy. Yet the pause, like silence, is directly relevant and keeps me faithful to my promise, of which this work is a result.

Keeping this promise is now all the more vital in view of my recent cancer diagnosis. The unanticipated has a way of reminding us of what really matters. For me, that is my undying love and devotion to Brooke, my bond with family and dear friends, and my loyalty to my students and to my profession. Brooke's faithful support, blessing, and presence continue beyond all measure. Not one word of this would have seen the light of day without her. To my precious Brooke, to whom I devote all my work, I remain forever grateful.

17 March 2015

Japan's March 2011 Disaster and Moral Grit

Our Inescapable In-between

Michael C. Brannigan

LEXINGTON BOOKS
Lanham • Boulder • New York • London

9月
30日
15年

Japan's March 2011 Disaster and Moral Grit

For Carolyn + Mac,
Thank you abundantly
for your steadfast
support + encouragement.
May you find this of
interest...
—Yours,
Michael

Moreover, it is increasingly the case that there are no *purely* natural, technological, or social disasters. All can be perversely intertwined. Japan's 3.11 is a case in point. The tsunami's breach of the seawall protecting the Fukushima Daiichi Nuclear Power Station can be attributed not only to the seawall's insufficient height, but also to lack of imagination in preparing for a worst-case scenario (demonstrated in placing back-up generators on lower levels) and systemic communicative *faux pas*.

Population

Characteristics of population, of all living beings not just humans, certainly factor into the scope of a disaster. Human demographics play a major role. The more heavily populated an area is, the increased likelihood of fatalities. The overall age of the population also matters. The more there are young children and elderly, the more there is risk of injury and death. Gender and gender roles also factor in. On March 11, all along the Sanriku coast, many Japanese mothers who generally stay at home to take care of their families died *en route* to pick up their children from school when the earthquake hit and tsunami ensued.

A population's primary means of subsistence is a crucial component. The March 24, 1989, *Exxon Valdez* oil spill in Prince William Sound destroyed surrounding natural resources that native communities depended upon for their livelihoods. Eyak, Chugach Eskimos, Native Tlingit and other cultures relied upon fishing and hunting throughout the area. And for fisherman and farmers along the Sanriku coast, whose work fed much of the Tōhoku (henceforward Tohoku) region, the tsunami and radiation from 3.11 has seriously handcuffed their future. Another key aspect of population lies in inhabitants' worldviews, beliefs, and values. These cultural dynamics inevitably affect preparation for and response to disaster. Tokyo's heavily urban culture contrasts with Tohoku's more rural lifestyle and outlooks, including emphases upon traditional rituals and festivals as sources of community strength.

Vulnerability

Vulnerability is usually construed in terms of physical vulnerability, that is, infrastructural development such as presence and expertise of emergency response, access to roads and evacuation routes and sites, condition of roads, reliability of power, water resources, and fuel, city planning, government plans and strategies, and hospital surge capacity plans. Each of these presents its own complex set of concerns. For instance, surge capacity in hospitals is not easy to precisely define. Generally, it refers to situations when, due to some sudden impact (like accidents or explosions) or unfolding crisis (such as a flu pandemic) that incurs present and potential large-scale injury, the dramatic excess, or surge of medical needs overpowers existing medical capabilities, for example, in emergency departments or intensive care units. According to the Joint Commission on Accreditation of Healthcare Organizations (JCAHO) early 2003 definition:

> Surge capacity encompasses potential patient beds, available space in which patients may be triaged, managed, vaccinated, decontaminated, or simply located; available personnel of all types; necessary medication, supplies, and equipment; and even the legal capacity to deliver health care under situations which exceed authorized capacity.[4]

Though specific, other considerations need to be factored in such as systemic communication dynamics, regional coordination, accessible transportation, and proper staff training. Proper hospital surge capacity is instructive when it comes to disaster response in general. As in any large-scale disaster, sufficient surge capacity preparedness requires integrating both the internal and external web of components involved in response. As some experts point out, "Critical points of failure exist when subcomponents are not synchronized. For example, the supplies and equipment must have the requisite and appropriately trained personnel to run the equipment."[5]

But there is more to vulnerability. Vulnerability also includes social vulnerability, that is, degrees of social and community cohesiveness and cooperation. This social vulnerability is a pervasive theme in this book as we explore expressions of community cohesiveness in the face of its dire absence, for example, for survivors still living in the many temporary shelters, or *kasetsu*, since 3.11. Although the definition below refers only to natural disasters, we can assert the same regarding others types of disasters as well, technological and human-incurred.

> By vulnerability we mean the characteristics of a person or group in terms of their capacity to anticipate, cope with, resist, and recover from the impact of a natural hazard. It involves a combination of factors that determine the degree to which someone's life and livelihood is put at risk by a discrete and identifiable event in nature or society.[6]

Later, I stress that this social vulnerability requires more than social linkage through digital communicative devices, or "connectivity." Since social vulnerability measures our degree of communal cohesiveness, this demands more than connectivity. It requires face-to-face personal interaction and engagement, what I think of as "connectedness" in the genuine sense. Ultimately, my subtext throughout this work concerns the underlying, crucial challenge: whether and to what degree we are "connected" *before* the disaster. For how we are with each other before the storm determines our capacity to survive and hopefully recover *after* the storm. Given that a disaster is more than simply the impact, unless we are fairly connected prior to the disaster, then our daily routine, our business as usual, can be in essence a disaster waiting to happen.

A Crack in the Discourse

Allow me to add to this conventional anatomy. Disaster occurs when what is routine scatters and, at least temporarily, breaks apart. In this splintering, fissures appear that reveal what is customarily hidden. These we can call, after philosopher-anthropologist-sociologist Bruno Latour, "cracks into the discourse," the "discourse" representing the routine.[7] To illustrate, consider face-to-face communication. Cracks regularly occur on this level. Discourse is the usual content of conversation, what Emanuel Levinas terms *le Dit*, what is said. There is a profound difference, however, between what is said as such, *le Dit*, and the saying of it, *le dire*. Cracks in the discourse of communication show themselves in the saying, *le dire*, how things are said, body language, and what is not said but still conveyed. Silence is the most powerful language, albeit the most difficult to learn.

Disaster, therefore, is a "crack in the discourse" of the routine and everydayness, a fissure in the banal, when what is taken for granted—water, food, clothes, mind, body, electricity, and shelter—takes on new meaning. Disaster experts Anthony Oliver-Smith and Susanna Hoffman write that "disaster often reveals the deeper social grammar of a people." At the same time, as they wisely remind us, this "deeper social grammar" is neither immutable nor timeless.[8]

Trauma, whether it is personal or collective, is a trial by fire, testing our survival skills. It tests our capacity to reach deeply within, to ascertain the genuine power of our inner resources. These inner resources often reveal deep-rooted, conditioned worldviews, values, and beliefs we may have "discarded" in keeping pace with the routine and banal. Trauma pricks the day-to-day routine and radically shifts, even shatters, the taken-for-granted. Indeed it sucks us into the marrow of the very conditions for the routine. Trauma has the capacity to awaken in us what really matters. This was most palpable in my encounters with numerous 3.11 survivors, victims' families, involved officials, recovery workers, and volunteers, all of which turned out to be an invaluable study "on the ground." In my conversations, they were precisely that, conversations, or discourse, in which I abandoned a formal interview approach and preferred to speak with victims, volunteers, and survivors on their own terms. Our communication was open-ended, allowing me brief glimpses into their worlds, and also through cracks in the regular discourse. One major crack in the discourse lies in being permitted into the interior, the inside of the survivors' temporary shelter, or *kasetsu*, zone. This was for me a profound privilege. As a "refugee in their own land,"[9] *kasetsu* became for each inhabitant his or her new *uchi*, a "home" away from home and, therefore, from any personal, private space of control, the outside representing loss of control, safety, and remains precarious. To be allowed inside the perimeters of their *uchi* was a generous opening, a crack in the conventional.

THREE CAVEATS

Before we proceed any further, here are three needed caveats. First, in view of my encounters with survivors and listening to their stories, we need to be cautious as to full literal accuracy of survivors' stories. Trauma displaces time and, in so doing, memory. Many survivors had difficulty relating what precisely occurred in those days following the impact that afternoon on March 11. Yet, they still sharply recalled the moments of the impact and what transpired that same day. It could be that they may have recalled hearing about the following events as they unfolded and took them to be their own, using others' stories as their stories, owning others' reports to fill in those gaps, in turn, satisfying a need for cohesiveness by restoring order after the chaos that trauma represents. Nonetheless, it is one thing to discuss a patient during rounds when residents, fellows, and students make their own calibrated assessments of the patient after their five-minute bedside visit, discussing the case from their own zone of safety and distance. So it is regarding narratives of others' suffering, the result of research ranging from what appears superficial to something in-depth, yet still engendered within a distanced safety zone. But there is no more authoritative voice than that of the sufferer, who lives the pain and whose being is determined by that pain. Try as we may, we who live outside their skin are always outsiders.

Here's another caveat. Be wary of "parachute professionals," those who jump into unknown territory and then become authoritative voices, "experts," regarding the territory. Like pundits and journalists, they are everywhere. And so with philosophers, notoriously noted for having both feet planted in the air. In fact, philosophers may be the most dangerous species of academics if they remain fixed under the tyranny of their own philosophic worldviews. There is a live tension between what comes down to mere exploiting and the need to share in and learn from others' suffering.

So, what's my excuse? Am I not one of those parachute philosophers jumping into the unknown to seek a story for my book? Throughout my experience and encounters in Japan after its 3.11 tragedy, I recorded reflections, experiences, and memories. However, trained as a philosopher and especially influenced both by Asian traditions, especially Japanese, and the rich and humane discipline of phenomenology, I remain committed to the notion that we philosophers have an urgent calling to somehow engage others via their shared lived experience. Furthermore, we as humans have a moral duty to do whatever we can to relieve another's suffering. I refuse to allow my professional role as an academic be an excuse for non-action, for a passive engagement caged in books and what others may write.

One of the freshest examples of commitment to engaged research and writing is the inspiring work of Renée Fox. She embodies devotion to her discipline, sociology, by way of lived experience. She artfully manages to do so without "othering" the Other, without exploiting subjects for her own narrative. Her recent *Doctors without Borders: Humanitarian Quests, Impossible Dreams of Médecins Sans Frontières* continues her vibrant and literate testimony to our calling as scholars to refuse to inhabit the world of spectator.[10] Thoroughly researched, it is the result of her immersion in key *Médecins Sans Frontières* sites, including Belgium, France, South Africa, and the Democratic Republic of Congo. Carolyn Nordstrom, University of Notre Dame anthropologist, is another testament to such engaged research. Her recent *Global Outlaws: Crime, Money, and Power in the Contemporary World* is a measured, honest, and morally sensitive product that illustrates a career of years of on-the-ground studies and immersion, placing herself at high risk in danger zones where she explores the complex, hidden world of international profiteering from crimes such as exploiting conflict minerals and human trafficking.[11]

In the past six years, my wife Brooke and I experienced our string of personal traumas when all four of our parents died, one after the other. Immediately after 3.11, I made a silent promise to my departed mother to learn more about the tragedy and share what I learn with others. So for me, disaster offers a necessary pause in my routine studies of medical and health-care ethics, intercultural studies, and philosophy. Yet the pause, like silence, is directly relevant, my own crack in the discourse that keeps me faithful to my promise.

Living in the spectator's village, the domain of silence, is easy, the path of least resistance, and bears little personal discomfort. It is far easier to remain in our comfort zones, our safe rooms. In doing so, we join the ranks of our most popular support group "Bystanders Anonymous," or what sociologist Stanley Cohen calls "Deniers Anonymous."[12] Collective silence spawns a collective denial. Stepping out of this represents refusal to be further complicit in others' suffering and, instead, pledges doing whatever one can, even at personal risk, to relieve others' suffering.

One final caveat. As always when it comes to studying cultures, we need to be extra cautious bequeathing generalizations. As I continually stress, when it comes to efforts to understand others in other cultures, no matter how well-intentioned we are, we must remain vigilant in avoiding hasty, stereotypical, and often ungrounded encapsulations. The first commandment in intercultural studies: thou shalt not cast monolithic nets over cultures.[13]

Alexandra Bech Gjørv, one of Norway's foremost respected lawyers, led the independent commission charged with investigating the July 22, 2011, summer camp massacre on the island of Utøya just outside of Oslo. Astute, articulate, and attentive, Alexandra's measured disposition and fair-mindedness made her the perfect leader for the commission. She shared with me various perspectives from camp survivors and recovery workers. The commission itself, a balanced mix of professionals and citizen representatives, thoroughly examined details of the human-incurred disaster perpetrated by lone gunman Anders Behring Breivik. Alexandra composed the final report regarding future security, response measures, and recommendations to prevent future instances.[14] I met with her on two separate occasions. Both times, she underscored the complexities of "culture" factoring into perspectives of the massacre. The commission's report focused more on the government's institutional culture and cited concerns regarding police efficiency, leadership, accessibility, and response time.[15] She described Norway's interweaving of various cultures and subcultures, loosely defined here as patterns of behavior: systemic, organizational, governmental, security, and leadership, all interfacing within a multicultural ethos increasingly embraced in Norway. All of these together present a complex labyrinth with unique sets of problems and challenges.

At the same time, this does not preclude assessments as to fundamental tendencies within cultures. For instance, Norway historically values trust, solidarity, respect, and equality. When the July 22, 2011, Utøya camp massacre occurred, the nation was stunned. I met twice with Anne Cecilie Fossum, Oslo's Red Cross director who played a key role in response to the disaster. We first met at an outdoor café close to the Oslo courthouse where Breivik was still being tried. The session had just finished and we were joined by Christin Bjelland, vice chair of the National Support Group, who also played a pivotal role in offering grief counseling and support to families and friends of the victims.[16] They both reaffirmed that Norway as a culture holds itself to the principles of trust, solidarity, and social democracy. It upholds the priority of collective well-being over individual self-interests. There were points during the ten grueling weeks of trial when this was apparent. One husband who lost his wife in the city bombing that took place just prior to the camp massacre took the stand to offer comments. He passionately appealed to the court to not respond to the tragedy in a way that would make Norway obsessed with national security, beefing up police security throughout the country. As Trondheim University psychiatrist Are Holen put it, to "keep Norways's innocence in this loss of innocence."[17] Anne Cecilie pointed out that social-cultural trust is evident in small things, when, for example, the King of Norway Harald V can go skiing in public without being surrounded by a cavalcade of security forces. And in response to the July 22 tragedy, King Harald spoke in terms of reconciliation, not retaliation or revenge.

As to Japan's earlier tragedy that same year, people who inhabit the region of Tohoku, where the 3.11 triple disaster unfolded, are not of one ilk. Those who live in Tohoku's three affected prefectures—Iwate, Miyagi, and Fukushima—cannot be branded or lumped together.[18] The people of Tohoku have uniquely different historical contexts, livelihoods, family links, education, worldviews, and lived experience. This is a profound proviso to bear in mind as we go forward and attend to some of their stories and listen to their voices.

Consider the region of Tohoku. We can mistakenly brand Tohoku as some compact geographic entity that embodies the "real Japan," having deep, immovable roots and traditions expressed via numerous village *matsuri*, or festivals. This may explain why, as anthropologist William Kelly points out, Tohoku has been the subject of much ethnographic and anthropological research.[19] Yet, regarding "culture," Tohoku is not homogenous. Its history is one of constant resettlement and change, particularly after World War II when a dwindling farming and agricultural lifestyle gave way to small industries, individual businesses and shops, with growing spurts of consumerism in towns like Tono, which is a merging of various towns, and a dramatic increase in an aging population while young people settle elsewhere for jobs. Post 3.11 recovery, therefore, is not simply a matter of rebuilding destroyed towns and villages. It requires what seems improbable, if not impossible—healing numerous factories fractured by the powerful earthquake; repairing a thriving fishing industry after waves overpowered fishermen, boats, processing plants and sea beds; and restoring a farming culture whose rice, vegetables, meat, and dairy products now hang under clouds of consumer suspicion. Who will return to live and work in devastated areas?[20] Moreover, each locale is situated within a larger and complex cultural, social, and political dynamic.

We can learn lessons from the sea when it comes to examining cultural values, particularly in island nations like Japan. Tides regularly shift. High tides only exist because there are low tides. For some time, Japan has been undergoing its low tide as long-standing, cherished, traditional values recede back into the sea, like victims swallowed by 3.11's waves. There has been less sense of relationality and secure and stable collective identity. There has been a diminishing bond between family, work, community, and national pride in such bond, as evidenced in growing skepticism in authority such as government and medicine. Despite the heavily ingrained impact of both Confucian and Buddhist thought on Japanese culture, notions regarding the primacy of relationships, which both traditions underscore, ought not to define our impressions of Japan. Think of traditional cultural views as the sea from which they come and that manifests itself in its tides. The difference, however, lies in predictability. Tides of cultural force appear irregularly. As anyone familiar with the sea knows, the sea does not conform to any ideal image. In the same way, culture simply does not fit into any fixed pattern or stereotype.

Despite this inescapable complexity regarding culture and Tohoku in particular, there are strong yearnings for a shared cultural and regional identity. After March 11, 2011, the figure and works of Miyazawa Kenji have emerged as a powerful cultural icon. Throughout my volunteer stint at Tono Magokoro Net, our group leaders have often alluded to his writings as an impetus for us to work together in rebuilding. Sudo, from Hokkaido, sometimes referred to Miyazawa, quoting him and describing his writings as embodying the spirit of the people of Tohoku. For my good friend Hiroko Suzuki and numerous others in Tono, Miyazawa represents an unbendable spirit of resilience in the face of adversity. In the many towns I visited, I have constantly seen reference to him. Read his poem, "Strong in the Rain," and you will understand why. It is a message of compassion, optimism, and resilience for the people of Tohoku. From my experiences and encounters, survivors in Tohoku indeed exude the spirit of resilience that Miyazawa here so humbly describes, persons who never lose their "generous spirit" nor fear death despite rain, wind, heat, cold, snow, and mortality. He speaks for many in Tohoku who claim "That is the sort of person I want to be."[21]

NOTES

1. Neil R. Britton, "Dog or Demon," in *What Is a Disaster? New Answers to Old Questions*, ed. Ronald W. Perry and E. L. Quarantelli (International Research Committee on Disasters: Xlibris Corporation, 2005), 113. This quandary about defining disaster is nicely discussed in Perry and Quarantelli's compilation, put together by two foremost experts in disaster studies.
2. K. Hewitt, *Regions of Risk. A Geographical Introduction to Disasters* (London: Addison, Wesley Longman, 1997).
3. Naomi Zack, *Ethics for Disaster* (Lanham, MD: Rowman & Littlefield Publishers, Inc., 2009), 7.
4. Joint Commission on Accreditation of Healthcare Organizations, *Healthcare at the Crossroads: Strategies for Creating and Sustaining Community-wide Emergency Preparedness Strategies* (Oakbrook Terrace, IL: Joint Commission on Accreditation of Healthcare Organizations, 2003), cited in Donna F. Barbisch and Kristi Koenig, "Understanding Surge Capacity: Essential Elements," *Academic Emergency Medicine* 13, no. 11 (November 2006): 1099.
5. Barbisch and Koenig, "Understanding Surge Capacity," 1011.
6. Piers Blaikie et al., *At Risk: Natural Hazards, People's Vulnerability, and Disasters*, 2nd ed. (London: Routledge, 2003), 11.
7. Bruno Latour, "How to Be Iconophilic in Art, Science, and Religion?" in *Picturing Science, Producing Art*, ed. Carolyn A. Jones and Peter Galison (London: Routledge, 1998), 429.
8. Anthony Oliver-Smith and Susanna M. Hoffman, "Why Anthropologists Should Study Disasters," in *Catastrophe and Culture: The Anthropology of Disaster*, ed. Anthony Oliver-Smith and Susana M. Hoffman (Santa Fe, New Mexico: School of American Research Press, 2002), 10.
9. David H. Slater, "Urgent Ethnography," in *Japan Copes with Calamity: Ethnographies of the Earthquake, Tsunami and Nuclear Disasters of March 2011*, 2nd ed., edited by Tom Gill, Brigitte Steger and David H. Slater (Bern, Switzerland: Peter Lang, 2013, 2014), 34.

10. Renée C. Fox, *Doctors without Borders: Humanitarian Quests, Impossible Dreams of Médecins Sans Frontières* (Baltimore, MD: Johns Hopkins University Press, 2014). Her work continues her rich tradition of engaged scholarship as in her earlier *Experiment Perilous: Physicians and Patients Facing the Unknown*, *Spare Parts: Organ Replacement in American Society*, and *Observing Bioethics*, the last two co-authored with Judith Swazey.

11. Carolyn Nordstrom, *Global Outlaws: Crime, Money, and Power in the Contemporary World* (Berkeley: University of California Press, 2007).

12. Stanley Cohen, *States of Denial: Knowing about Atrocities and Suffering* (Cambridge, UK: Polity Press, Blackwell Publishing, 2001), 256.

13. I've stressed this caveat about "culture" particularly in my *Healthcare Ethics in a Diverse Society*, co-authored with Judith A. Boss (Mountain View, CA: 2001), *Striking a Balance: A Primer in Traditional Asian Values*, rev. ed. (Lanham, MD: Lexington Books, Rowman & Littlefield Publishers, Inc., 2010), and *Cultural Fault Lines in Health Care: Reflections on Cultural Competency* (Lanham, MD: Lexington Books, Rowman & Littlefield Publishers, Inc., 2012).

14. Norges offentlige utredninger 2012: 14, *Rapport fra 22. Juli-kommisjonen* (Oslo: Departementenes servicessenter Informasjonforvaltning, 2012) [Norwegian Public Reports 2012: 14, *Report from The 22. July Commission* (Oslo: Government Service Center Information Management, 2012)].

15. For example, just one day after the report came out, the chief of police resigned. Doubts regarding his proper qualifications arose during the investigation into police response following the 7.22 massacre. Though a well-respected physician, some questioned his background in his position of leading the Oslo Police. Since his resignation, he now heads one of the hospitals. From an interview with Alexandra Bech Gjørv, May 12, 2014.

16. Interview with Anne Cecilie Fossum, Oslo, June 21, 2012.

17. Interview with Are Holen, Oslo, June 20, 2012.

18. A prefecture, or *ken*, is second to the federal government in administrative authority. Prefectures are similar to states in the United States.

19. William W. Kelly, "Tohoku's Future: Predicting Outcomes or Imagining Possibilities?" *The Asian-Pacific Journal: Japan Focus*, accessed Feb. 5, 2015, http://japanfocus.org/-William_W_-Kelly/3703.

20. Kelly, "Tohoku's Future," http://japanfocus.org/-William_W_-Kelly/3703.

21. See http://mooninthewater.net/aikido/2012/10/13/translating-japanese-strong-in-the-rain/, accessed Jan. 10, 2015. I was just diagnosed with transitional cell upper tract urothelial carcinoma in my kidney just before writing this. A small trauma for me, bigger for Brooke. His poem for me hits home.

Chapter One

Unhinged

Nature has its unforgiving way of reminding us that we are inescapably in-between, situated within time and place. On Friday, March 11, at 2:46 p.m., a Goliath-like trembling, later measured at a magnitude 9 earthquake, erupted seventy kilometers, 43.5 miles, off Japan's northeast Sanriku coast. There, the earth snapped at a sea-depth of eighteen miles. The underlying tectonic plates' shift violently lifted the seafloor upward, generating enormous tremors that cracked and rattled the earth throughout a good portion of the large northeast region of Tohoku and along the Sanriku coastline. Japan's Pacific coastline dropped 1.2 meters (3.9 feet) and moved Honshu, Japan's principal island, eastward by 5.3 meters, or 8 feet. Japan is now one parking space closer to our West Coast.[1]

How can something that deep in the ocean generate such force? Our planet, misnamed "earth" since it is two-thirds water, has a stunningly incomparable life of its own given enormous collateral damage from just subtle shifts of tectonic plates. The tectonic plates moved with enough friction to release an explosive energy of one million kilotons of TNT. The energy released at Hiroshima was fifteen kilotons.[2]

This is the island archipelago's largest recorded earthquake, or *jishin*, outsizing both the Great Kanto Earthquake of 1923 that measured 8.3, and the more recent 1995 Kobe, or Great Hanshin's, 7.2 earthquake. Everyone throughout Japan's main island of Honshu and far beyond felt the ground's prolonged convulsions. Japanese are used to tremors. But this one was different. This was unlike other shocks. There was no sudden jolt. Instead, there was a slow, undulating, increasingly powerful trembling that lasted for six never-ending minutes. Throughout the northeast region, violent shuddering lingered endlessly. The ground swayed nonstop, buildings shuddered, shelves crashed to heaving floors, structures came apart, buildings cracked, and roads ripped open.

All became unhinged. The unraveling was especially profound in crowded Tokyo, 373 kilometers, or 231 miles, southwest of the epicenter. There, lights waved frantically, trains immediately came to a standstill, all public transport stopped, businesses and schools halted, and darkness suddenly descended as all power went out. In metropolitan Tokyo, people were confined and millions of buildings were without power. On that day, March 11, there were 23 aftershocks with magnitudes ranging from 6 to 7.9. Days later, once power was restored, those who could afford it embarked on a mass exodus away from the city heading southward to places like Osaka and Kyoto, and out of the country to Hong Kong, Thailand, and South Korea.

Mizuki Naito, now recently married as Mizuki Abe, was working at her desk in the East Japan Railway Culture Foundation headquarters in the heart of the city, near Shinjuku Station. Mizuki is a top-level assistant to the Culture Foundation manager Kunio Aoki. She and Kunio later accompanied me to Tohoku during my visit and research there. Once the rumbling started, Mizuki knew this was out of the ordinary. She and her coworkers were ordered to stay inside the building and not venture outside due to the danger of falling debris. They were later required to stay overnight. Offices throughout the city did the same, instructing employees to remain indoors. Mizuki spent the night along with her coworkers sleeping on the floor. There is good reason to not venture outdoors. That day, around 268 people suffered fatalities. Many of them were crushed to death from falling objects. Unlike Hollywood depictions of earthquakes where the ground swallows its victims, death comes from above. Death also comes from fires which almost always accompany earthquakes. And they spread rapidly with hazards like the oil refinery in Ichihara, close to the city. The refinery soon became an inferno. The quake generated fires and landslides that claimed around 165 lives.[3] Compared to Kobe's earthquake on January 17, 1995, these figures were low. That one took over 6,400 lives. But Kobe's quaking damage was concentrated in Kobe. Tokyo's fracturing had its source from an epicenter over two hundred miles away.

Mizuki had no way to contact her mother. Her mother lived alone on the outskirts of the huge city. Power lines and mobile phone access were shut down. Trains stopped. There was no public transport. After workers were allowed to go home the next day, she ventured outside to journey to her mother's to ensure that she was safe. Convenience stores, a familiar sight throughout Japan, ran out of milk and water. Mizuki stopped at the few convenience stores along the way that were open. Yet they had already sold out their supply of milk, water, and batteries. These were the first to go once yesterday's shuddering subsided. In the days ahead, even umbrellas, popular during this rainy season, were in short supply. This was not due to excessive rainfall. Soon after power was restored in Tokyo, news centered 24/7 around the terrible unfolding tragedy up north, the massive tsunami and the damaged nuclear reactors. With all eyes fixed on the disaster, rumors sprouted about rain being contaminated with radiation from the Fukushima Daiichi Nuclear Power Station. The reports were not entirely groundless. Heightened radiation levels were detected in universities and hospitals. To make matters worse, however, in the midst of growing fears, inconsistent information from TEPCO (Tokyo Electric Power Company), government officials, and radiation experts in the industry and in universities contributed to the spread of unfounded speculations.

Within a short time, some trains began operating. Mizuki was able to board one to get to her mother. Apart from nerves, her mother was fine. Along the way, however, Mizuki was shocked to see the dire shortage of gasoline, especially affecting the hundreds of taxis lined up for hours at filling stations. Once power was restored, millions of Tokyoites resumed work. There was no rioting, looting, or public disorder apart from some glaring tempers from drivers waiting for hours to buy gas.

The world still wonders at the absence of riots and looting, of the sort that occurred in the aftermath of disasters in places like Katrina and Haiti. As we will see, much of this has to do with a heightened awareness among Japanese of our being naturally and unavoidably situated in-between both natural climate and human climate. This means that how we respond to the whims of our natural climate, its weather and natural hazards, depends very much on how we relate to and with each other in our human climate. And vice versa. Maintaining human order and decorum is thereby a crucial prerequisite to survival. However, let us avoid hasty generalizations. Other factors help explain this absence of disorder. In mega-metropolitan Tokyo, with a population well over 13 million and in many respects unlike rural Tohoku, the general public was not that well informed as to the situation's seriousness at the power plant. Two months later, in May, mainstream media finally informed the public of the meltdown inside reactors one, two, and three, with reactor 4 in serious condition. Nonetheless, the pressing need to sustain human order, decorum, and overall concern for the common good is evident in that after power was restored and well after the government's compulsory period of limitations on power usage, businesses still voluntarily maintained conservation measures.

SECONDS COUNT

> As soon as they had been brought outside, he [Lot] was told: "Flee for your life! Don't look back or stop anywhere on the Plain. Get off to the hills at once, or you will be swept away." (Genesis 19:17) . . . As instructed by the angels, Lot, his wife, and their two daughters fled towards the hills. But Lot's wife looked back, and she was turned into a pillar of salt. (Genesis 19-26)[4]

Further north, around forty minutes after the intense tremors, waves reached the coastline, towering up to 12 meters high, over 39 feet, and higher in some spots, lunging inland as far as ten kilometers, or six miles. Imagine that the last thing you see is an oncoming black wall of mud, debris, oil, gasoline, and bodies. For countless victims along the Sanriku coast, the sea, the color of lead, hit them like steel with the force of a powerful locomotive. Those who made it, like Jin Imai and Bunzo Kanayama, whom we will meet later, recall hearing the roar of the wave sounding like a train.

A tsunami following an earthquake is to be expected. Sanriku is not new to tsunamis, and many towns hold annual tsunami evacuation drills. Seawalls, around six to seven meters high, or twenty feet, protect much of the coastal region. And Japan's sophisticated earthquake and tsunami alarm system is second to none. Yet, the Japan Meteorological Agency vastly underestimated the tsunami's size and speed. Most importantly, it failed to consider the worst-case scenario. The highest warning category was set for a tsunami three meters (9.8 feet) in height.[5]

Because we live helplessly in-between time and place, timing is everything. So is place. For all who were there along the coast, this was the worst time. Families were split up at work, school, home, and on the road. Because of Tohoku's mountainous topography in that area, towns are scattered and spread out. The waves' mid-afternoon timing decimated local leadership at town halls where many died at their posts. Like twenty-five-year-old Miki Endo who continued to publically broadcast warnings to citizens in Mimamisanriku, one of the worst hit towns, to head to high ground. Her last moments were spent urging everyone to escape until the giant wave swept her away. Her broadcasts surely saved thousands of lives. All along the coast, ordinary citizens responded as trained, seeking higher ground and heading towards designated evacuation shelters. However, evacuation shelters at gyms, community centers, elementary schools, and temples became death traps unless they were situated on much higher ground. Many villagers in Otsuchi flocked to the designated Buddhist temple for shelter. Yet, the force of this wave was far too powerful, crushing everyone there, nearly one hundred people. For those who headed upstairs in their homes, they may well have stayed on the ground floor. The waves consumed them.

Seconds count. Many made the fatal mistake of looking back. Many others sought to retrieve valuable household items. But, like Lot's wife who was killed for violating the angels' warning, villagers died for turning back. Doing so, to even look back for a split second, meant death. Jin Imai, working with his crew on the outskirts of Otsuchi, literally sprinted for his life up the hilly road. His brain told him—"Don't look back. Not for one second." His coworkers did. They never made it. Onrushing waves with their formidable speed and force swallowed cars that took wrong turns, including school busses filled with children.

Giant waves literally battered and buried the northeast coastline. For weeks afterwards, the coast was a cesspool of human carnage, debris, household items, splintered boats, cars, sharp metal, jagged glass, farmland, trees, bushes, broken timber, all buried in black mud. Another heartbreak in this litany of tragedy are the countless dying, dead, and lost animals. Despite efforts from Animal Rescue in Kansai (ARK) and Japan Earthquake Animal Rescue and Support (JEARS), the priority was to save as many humans as possible. Thousands of pets and farm animals were abandoned, left for dead, and missing. For months afterwards, the lingering stench of death pervaded the coastline.

Countless numbers of recovered bodies littered the scene, covered in blue vinyl sheets. Yet, there were thousands still missing. This adds further insult to unimaginable suffering. It is especially traumatic when a body is never found. Without a body, there is little closure on the part of family and friends. Grieving becomes harmfully prolonged, often hidden. Without the body, a proper memorial service is extremely difficult. Re-possessing the body, no matter how disfigured, is crucial for survivors. For those not immediately killed by the waves, chances of surviving were bleak. They faced the ubiquitous debris, incessant cold, winds, snow, rain, blazing fires, absence of power and clean water, impassable streets and roads, and the darkness. As many survivors refused to leave what was left of their homes, waiting for relatives who never came, and for others huddled in shelters, waiting for help and hanging on to hope, they faced the flood of total darkness.

Think the Unthinkable

Further south along the Sanriku coastline, another disaster was unfolding, something even more insidious, if we can even fathom anything surpassing the tsunami's utter devastation. This was the unraveling of a natural disaster amidst the heavily technological climate of nuclear power. Here, victims not only faced the earthquake and waves, but a third, invisible enemy—radiation contamination.

Minutes after the giant earthquake, the massive tsunami swarmed easily over seawalls protecting the Fukushima Daiichi Nuclear Power Station, owned and operated by Tokyo Electric Power Company (TEPCO). Waves easily drowned the reactors' back-up cooling systems. In the days ahead, fuel rods in at least three of the six reactors reached temperatures over two thousand degrees Fahrenheit causing them to melt down. This prompted radioactive material to penetrate containment vessels and spew forth into the soil, groundwater, and sea. Moreover, incredibly intense heat triggered hydrogen explosions of the tops of three reactors releasing toxic gas and radioactivity into the atmosphere.

This event in particular starkly illustrates what is perhaps our deepest-rooted human flaw—our built-in incapacity, or unwillingness, to think the unthinkable, to conceptualize the unimaginable. Along these lines, Freud remarked that we cannot conceive of our individual non-being. This constitutes an inherent feature of human vulnerability, and collectively this forms part of our social vulnerability. This limit to our mental stretching may help explain why millions along our U.S. Pacific Western coastline go about their businesses, knowing that someday soon the Big One will arrive. Or why officials in Tokyo, despite predictions of a 70 percent likelihood of a magnitude 7 earthquake occurring in Tokyo by 2016, are still moving ahead full-throttle in preparing for the 2020 Summer Olympics.

One major reason why the Daiichi facility was not well-prepared was that there was and continues to be little consideration given to factoring in worst case scenarios when calculating and assessing risks. It comes down to refusing to learn from history, deliberately ignoring lessons from the past. Four years earlier, in 2007, a 6.8 magnitude quake hit close to one of the world's largest nuclear power plants at Kashiwazaki-Kariwa, located on Japan's west coast in Niigata Prefecture facing the Sea of Japan. The quake's large tremors caused sizable damage to monitors, transformers, and interior chambers. Nature delivered her wake-up call for imagining the worst case scenario—total power failure including back-up generators—precisely what happened on March 11, 2011. Without power, water that would normally cool fuel rods would eventually dissipate and overheat, and fuel rods would melt. Melting fuel rods would release deadly radiation, indeed what happened days following 3.11. Yet, at the Daiichi plant, preventive measures were not put into place to prepare for this worst-case scenario of the complete absence of power. Within days, not only were three reactors nearly completely damaged, but over one thousand fuel rods in reactor four became bone-dry. The unimaginable could unfold—a nuclear fission chain reaction.[6] Such a nightmare chain reaction could spark meltdowns at nearby plants like Fukushima Daini, Daiichi's sister plant just seven miles south, as well as other power plants along the Sanriku coast.[7]

As it turned out, though huge waves hit the Daini plant and, combined with the king-size tremblor, incurred heavy damage, almost miraculously the plant did not suffer a core meltdown. Whereas the Daiichi plant lost total power and workers were completely in the dark, at Daini a single generator was left intact and there was at least some electricity. Due to steady leadership and knowledge of the site on the part of the Daini's superintendent, Naohiro Masuda, he managed to maintain focus among his four hundred employees. And within days after 3.11, with assistance from TEPCO and Self Defense Forces (SDF), workers were able to connect over 9 kilometers (over 5 miles) of cable weighing over forty-five tons to the one working generator and prevent reactor cores from overheating and melting down.[8] The survival of the Daini plant offers a telling lesson in crisis leadership.

There is a deep layer of difference between the earthquake/tsunami disaster, or natural disaster, and the nuclear meltdown, or technological and human-incurred disaster. After waves took thousands of lives in the coastal area near the reactor plant, hundreds of thousands more in the vicinity of the nuclear plant were forced to move out of their homes. These displaced victims face a decidedly different sort of grim threat than do the homeless in temporary shelters further up north. They face the unseen threat of radiation, along with its fraternal twin, uncertainty. The menace of radiation poisoning along with future uncertainty coexist to imperil families, their children, and their children's children for generations to come.

JAPAN'S IN-BETWEEN

Japan's physical in-betweenness is clearly evident in its geography, topography, and geology that makes it one of the world's most disaster-prone countries. The country sits atop active volcanic and seismic zones. Its situatedness makes Japan especially susceptible to earthquakes, tsunamis, typhoons, severe rains, and volcanic eruptions.

The country's geography, topography, and geology naturally affect its climate, its overall weather pattern. Japan's climate undergoes extreme variations, from oppressive heat and humidity in the south to blistering cold and snow in the north. Its rugged mountainous topography sits on shifting tectonic plates, thus subject to constant earthquakes. The country undergoes approximately 1,500 minor tremors annually. In 2003, there were over two thousand recorded earthquakes. Japan's isolation as an island surrounded by sea makes it especially vulnerable to tsunamis and fierce typhoons. Typhoons come around every summer and can produce devastating landslides and floods. In September 1959, in one of the worst natural disasters in recent decades, the Ise Bay Typhoon took over five thousand lives.

Along with earthquakes and tsunamis, around fifteen volcanic events occur yearly. The Japanese archipelago is part of a mountain range known as the Circum-Pacific Ring of Fire, due to the Ring's numerous volcanoes. The Ring harbors around 75 percent of the world's active volcanoes. Japan itself has 7 percent of these. The damage from erupting volcanoes is another one of Nature's screams. On July 15, 1888, the rampaging, all-consuming avalanche from the explosive eruption of the volcano Bandai-san, northwest of Tokyo, smothered around five hundred villagers in seconds. In fact, a photo capturing the splitting apart of Bandai-san in the newspaper *Yomiuri Shimbun* was the first photo in a Japanese newspaper.[9] Of its 450 volcanoes, experts at the Japan Meteorological Agency estimate that 110, including the sacred Mount Fuji that hundreds of thousands of people climb every year, are active.[10] Currently, for a volcano to be considered "active," this means that it has erupted in the last 10,000 years. This seems incredibly long by our standards, but is a brief moment within the context of the Earth's 4.6 billion years.[11] Japan recently suffered its worst postwar volcanic eruption with the outburst of Honshu's Mount Ontake on September 27, 2014. Up until 1979, Mount Ontake was actually considered "extinct" until it erupted that year. Mount Ontake, the second highest volcano in Japan, is around 120 miles southwest of Tokyo. Over fifty hikers lost their lives. Although the eruption was a complete surprise, not at all predicted, it may not have been completely random. The March 11 earthquake was massive. And a string of constant earthquakes that naturally stimulate further crustal movements can lead to volcanic eruption.

A Culture of Disaster Preparedness

Because Japan lives daily with the imminent threat of natural disaster, disaster-preparedness is built into its society and culture. Companies and schools throughout Japan hold regular earthquake drills for employees and students. Ever since 1960, the first week in September has been its annual National Disaster Prevention Week. September 1 is National Disaster Prevention Day, and kicks off the week with a national mandatory earthquake drill simulating a real disaster.

This commemorates September 1, 1923's 7.9 magnitude Great Kantō Earthquake that took over 142,000 lives. It obliterated Tokyo and Yokohama. Timing, again, is everything. The earthquake occurred during lunchtime, when many were cooking with lit stoves. The jolts caused numerous fires, burning nearly half a million wooden homes to the ground. To learn the lessons from that tragedy, this entire first week in September is one of continuing education for the community regarding earthquakes, tsunamis, and typhoons, Japan's most frequent and anticipated hazards.

Two related questions. What led to the year 1960 as the start of Japan's annual Natural Disaster Prevention Week? And who would believe that an earthquake that occurred halfway around the world would lethally impact them? On May 22, 1960, at 9:11 a.m. Hawaii Time, Chile encountered the largest earthquake in recorded modern history, measuring in at 9.5 magnitude. The ensuing tsunami with waves eighty feet high devastated the Chilean coastline, literally taking out the city of Valdivia. Nearly six thousand people were killed. Fifteen hours later, around midnight, the tsunami hit Hawaii and pulverized a good portion of the island of Hilo, taking over 60 lives.[12] The tsunami reached Japan seven hours later, killing over 140 people. This was twenty-two hours after the earthquake! Though the waves were relatively small in most places by Japanese standards, around ten feet, at other spots they reached twenty-five feet. The Japan Meteorological Agency (JMA), not having had this experience from an earthquake so far away, "halfway around the world" at 10,000 miles, understandably underestimated the size of the tsunami. This inspired Japan, along with other nations, to centralize a warning system for that region, later known as the Pacific Tsunami Warning Center.[13] Until 1960, the scope for JMA's earthquake and tsunami prediction was local, not global. Japan had more than its fair share of local earthquakes and tsunamis. With the Chilean earthquake, Japan was brutally reminded of its inescapable in-betweenness, its *aidagara* that permeates all of us. The Chilean earthquake and tsunami left a deadly footprint of our unavoidable interconnectedness.

All this reminds the country that natural disasters are to be expected, and the key to surviving lies in being informed and well-prepared. A natural disaster can occur anytime. No part of the country is immune. The secret to survival lies in education and preparation. Here is Japan's built-in lifeguard mind-set. This is particularly evident in Japan's earthquake detection and warning system.[14] It remains unparalleled, second to no other in the world, and its warning infrastructure is by all measures first class.

The country's earthquake early warning system, *Kinkyu Jishin Sokuho*, offers an early alert as to estimated magnitude, seismic intensity, arrival time, and expected location. Magnitude measures the size of the earthquake, that is, the energy released at the epicenter. It is measured quantitatively, for example magnitude 4, or 7, measured through seismographs. It operates according to a logarithmic scale, which means that an increase of one on the magnitude scale (for example, from 4 to 5) amounts to increased energy by ten times. A magnitude 3 shakes ten times more violently than a magnitude 2.[15] The Japan Meteorological Agency distributes around two hundred seismographs and six hundred seismic intensity meters throughout the country. In addition, it obtains further data from an additional 3,600 seismic intensity meters operated by local governments.

These seismic intensity meters measure ground motion at specific sites which will test the locale's geology and topography as well as distance from the epicenter. Seismic intensity is more qualitative, more subjective, in that it has to do with the degree of shaking in *specific locations* and thus varies. I may be awakened by a slight tremor, but friends who live, say, twenty miles south of me may feel more shaking. Intensity pertains to location relative to the epicenter, geography, geology, topography, human-built physical structures (for example living in a wooden, concrete, or brick house), physical infrastructure, and so forth.[16] In other words, the experience of an earthquake's intensity rests upon our natural situatedness or in-between.

Knowing all this ahead of time helps to minimize earthquake-related damage and injuries. This is historically evident in Japan after the Meiji Restoration (1868–1912) commenced in Japan. The more the science of seismology developed, the more this affected Japanese architecture and increasingly sophisticated structures that became more earthquake-resistant.

Seismographs and seismic intensity data will also immediately convey the likelihood and size of tsunamis. If a tsunami is imminent, within minutes following the tremors, the JMA issues a tsunami warning to the coastal regions most likely affected. While mechanisms now in place such as the Seismic Sea Wave Warning System can better predict the timing of a tsunami, predicting its size remains difficult. Nonetheless, warnings, when heeded, have certainly saved lives. The operative word here is "heeded." On March 9, just two days before the day of reckoning, a tsunami alert followed on the heels of violent tremors along the Tohoku coastline. Yet this generally went unheeded. People who live by the coast have heard it before, and, in this case, familiarity spawns indifference. This mind-set was the game-changer between living and dying on March 11, 2011.

When warnings are unheeded, casualties inevitably result. Again, we are all indelibly in-between time and place. Consider what happened on the Hawaiian island of Hilo as a result of the 1960 Chile earthquake. Chile's earthquake, at 9.5 magnitude, the largest recorded quake in modern history, wracked untold destruction along Chile's coastline. Fifteen hours later, despite accurate prediction of its arrival time in Hawaii, many citizens of Hilo downplayed the warning. Why? First, they have had a recent history of false positives. Next, there was misunderstanding regarding the new warning system. Third, when media broadcasts described the diminutive force of the waves striking Tahiti hours earlier, many in Hilo assumed they were safe. This was a fatal mistake. Coral reefs surround Tahiti and therefore act as a natural barrier. In addition, Tahiti lacks a continental shelf which would have increased wave size. A tsunami wave slows down and increases in height when it reaches more shallow waters. Hilo citizens and media did not take into account Tahiti's natural protection from barrier reefs and its absence of a continental shelf, the kind of natural protection Hilo did not have. Tsunamis generate a series of waves, often called a “tsunami wave train.” So, even though the initial wave was relatively small, the unexpected third wave, a thirty-five-foot killer, virtually demolished the city, drowning 61 people and injuring over 280 others.[17] Lesson—not all islands are the same. We all live in our own unique in-between.

AFTERMATH

Cold as Hell

What do Hannibal, Napoleon, and Hitler have in common? Their invading forces were brutally decimated not by their enemies, but by the weather—the bitter, stinging cold from winter’s four-part harmony of wind, rain, snow, and ice. There is a reason why Dante’s lowest circles of Hell were the coldest. Hell on earth can be the coldest spots. When he was a young soldier, my mother’s older brother died in a prisoner-of-war camp in Siberia. With temperatures often crashing down below -40 degrees Celsius, exposure did him in. Those who survive the cold in places like Siberia, Northern Russia, Himalayas, and the Antarctic do so only if they have shelter, warmth, dryness, and nourishment.[18]

Many survivors, in shock for hours immediately after 3.11’s sudden impact, lacked all of these. Drenched to the marrow in freezing temperatures, they only had what they wore. Subjected to wickedly harsh and unpredictable wind and freezing rain, many, especially elderly, most likely soon died of exposure.

Remember the ill-fated Air Florida Flight 90? On January 13, 1982, it crashed into Washington, DC’s 14th Street bridge and plunged into the icy Potomac River following takeoff from Washington’s National Airport. 73 passengers died along with four motorists. Only six survived. Just the shock itself upon impact with cold water can cause a fatal arrhythmia. The victim can also hyperventilate from seizure of the muscles, and then become unconscious and drown. Immersion in freezing water induces hypothermia within 15 minutes and death in 30 to 90 minutes. Most likely, more fatalities from the sinking of the Titanic occurred from hyperthermia than drowning. And, when victims tend to panic in the cold water, struggling only makes it worse, speeding up heat loss and more quickly inducing hypothermia.[19]

Again, timing is everything, and this was the worst of times weather-wise in Tohoku, particularly along the coastline. This is the time of incessant wind, wet, and cold. Here, wind-chill is literally that, when the wind can speed up in moments for anything to freeze. And the wind need not be strong or gale-like. A wispy ten mile an hour wind can drop the air temperature around 20 degrees.[20] Cold, rain, ice, and snow all combine to induce bodily heat loss. Such heat loss will eventually get to the point when the body becomes so numb with cold, it cannot feel hypothermia kicking in. Hypothermia occurs when heat loss exceeds heat gained, or more clinically, when our core body temperature (measured deep within our chest and abdomen) drops below our normal 35 degrees Celsius.[21] Signs of hypothermia are first evident with violent shivering. This is moderate hypothermia. Once our core drops below 32 degrees, our bodies stop shivering and simply shut down. Survivors faced nonstop cold. And the ensuing cold adversely affects nerves and dexterity. Increased exposure to the cold for long periods brings about excessive loss of bodily fluid loss, dehydration. Moreover, the colder it gets, the more food we need, more calories for our bodies to generate enough heat to keep up with the cold. Hunger accompanies hypothermia. On 3.11, those who were not killed by the waves immediately faced the risk of severe hypothermia and freezing to death. They needed shelter, warm clothing, blankets, food, and water. For days after impact, these simply did not exist.

Immediately after this triple-horror, an estimated 700,000 survivors were squeezed into the few remaining evacuation shelters and other intact buildings such as community centers and schools that were not swept away by waves. There, they eked out each day without power, running water, warmth, and clothes. With splintered roads, ruined railway lines, mud, debris, and dead bodies everywhere, delivering blankets, food, water, clothes, and umbrellas during this rainy season was especially daunting. During this bitter cold season, survivors especially needed coats to stay warm. What's striking is that distributing them fairly did not rest upon a lottery system or hierarchy of needs, as was the case in later assigning people from these evacuation shelters into temporary dwellings, or *kasetsu*. It was often the case that if a coat could not be provided for all survivors in an evacuation center, crammed to the gills, then they could not be distributed. It was a matter of all or none. Nonetheless, people from all over Japan and the world donated numerous items. Yet due to the system of distribution, many of these items in the form of clothes were left unused. They were dispersed elsewhere. Our volunteer center Magakoro Net base camp in Tono held many items—coats, hats, gloves, blankets, boots that were originally intended for survivors. They were hanging in the hallway beside our meeting room. Some of these were later distributed to survivors in various *kasetsu*. Some were used by volunteers themselves, particularly during the harsh cold season when I was there in 2014.

In any case, soon after impact, the government stepped up to the task and responded quickly, sending in 100,000 Self Defense Forces (SDF) troops, one-half of the country's entire force. The official response here was radically different from the response to Kobe's earthquake on January 17, 1995. The Great Hanshin quake claimed over 6,400 victims. Yet aid from the government came slow, and the SDF's involvement was minimal. In this case, their help was crucial. These troops were charged with the dangerous and abrasive task of search, rescue, recovery, clearing power lines and roads, placing survivors into evacuation shelters, providing water, food, clothing, supplies, and endlessly sorting and hauling away debris.

Bodies Found and Missing

Countless bodies covered the muddy landscape. SDF troops faced the gritty task of recovering, unearthing, and gathering together bodies and body parts. They were not the only ones. Survivors conducted their own search and recovery of family members and loved ones.

Finding their deceased posed a double trauma for family members. It was bad enough to find and report one's parent, spouse, child, sibling, or other family member as dead. In many instances surviving family then had to retrieve a dry plank upon which they could place their loved one. They then needed to find a blanket, wrap the body in the blanket, and somehow manage a way to carry it on the plank to an area to which they'd been assigned. For family members, this was nearly impossible. There was certainly no shortage of bodies. However, dry wood and usable blankets were in exceedingly short supply. Waves had decimated entire villages. Everything was muddied and drenched. And finding a way to transport the body was exceptionally tricky given the lack of transportation and dangerous debris. Nonetheless, for weeks following impact, recovery workers, SDF, survivors, and family members retrieved and dug up thousands of bodies buried in mud and under debris, trapped in buildings, and washed ashore. SDF troops placed these bodies in schools, tents, community centers, and warehouses so that survivors could come and identify them. Soon, however, for reasons of health, bodies both identified and unidentified needed to be disposed of, at least temporarily. The customary cremation was not much of an option.

After the Pacific War cremation became more of the norm, replacing the tradition of burial. The combination of increased population, mounting urbanization, and with it growing scarcity of usable land led to the now prevailing custom of cremation. After March 11, however, mass graves were unavoidable. There was little fuel and wood for cremation. Bodies were thus placed together in temporary mass burial sites, separating the identified and unidentified. Once there were enough materials and services for cremation, they were later disinterred. The tsunami devastated entire towns. Throughout Sanriku, this meant that the usual resources used for cremation were truly scarce. To begin with, many undertakers who handled the complex process of cremation were killed or missing. Next, there were little means to transport found and identified bodies to the few remaining cremation centers that were not demolished by the earthquake and waves. And even if a vehicle could be found to carry the bodies, gasoline was in short supply. Most places had already run out of gasoline and were awaiting fuel trucks from outside the area. But because the earthquake had destroyed passageways through the few routes to the coast, and due to fears of radiation further south in Fukushima Prefecture, many fuel truck drivers chose to neither traverse the dangerous roads nor venture into areas that may be poisoned through radiation.

Cremation is particularly vital in that it serves a role in sustaining a sense of bodily integrity or wholeness. And the importance of maintaining bodily integrity plays a crucial role in Japanese culture. With the advance of medical technologies, life-sustaining treatment, and organ transplantation, controversies first arose in Japan over transplanting bodily organs. Procuring another organ meant disrupting the integrity of the body of both donor and recipient. Nonetheless, kidney, corneal, and eventually liver transplants were condoned for reasons of compassion, saving the patient, and enabling donors to manifest their altruism. Heart transplantation was another matter. There was a long-standing and bitter dispute over the moral legitimacy of transplanting a heart. Among the reasons for opposing heart transplantation were concerns regarding disrupting bodily integrity. Other reasons, however, were more prominent.[22] Eventually, due to a variety of factors including public pressure, pressure from professional groups like the Japan Medical Association, national embarrassment over having its citizens seek heart transplants elsewhere, and the supreme Buddhist importance of the virtue of compassion, the Japanese Diet passed a heart transplant statute in 1997.

Bodily integrity remains a serious matter regarding corpses. Corpses are not simply viewed as dead bodies, but bodies that still possess the spirit of the former person. Corpses are therefore to be treated with the utmost respect. SDF troops demonstrated this respect for bodies throughout their recovery ordeal. Working in twos and threes, they often spread out throughout the stricken area, trudging very carefully in the mud to avoid dangerous and lethal debris. When they came upon a dead body, its hand, foot, or face peering out from the mud, or found body parts, they uncovered the body or body part, placed it in a body bag, and then stood over it and offered prayers, hands clasped in sincere, natural reverence. One traditional reason for treating the dead body with respect lies in properly preparing the deceased for crossing into the "Pure Land" Buddhist paradise or spirit world of the gods, *kami*. And while Japan has become saturated with modern technologies, this traditional belief still remains strong in more rural areas.

Tohoku, 3.11's victimized region, is one of these areas. Tohoku still retains many traditional beliefs, rites, and practices, particularly with respect to handling the dead body. Long-standing Shinto and Buddhist beliefs prevail throughout Tohoku, and this holds true along the Sanriku coast. In major cities like Tokyo and Osaka, preparing the corpse for cremation takes places in hospitals, with the work often done by nurses. Yet in Tohoku, the mortician is the one who carries out the cleaning, preparing, and dressing the body before placing it in a coffin for viewing. The mortician, or *nōkanshi*, performs this crucially important ritual of cleaning, preparing, and dressing, called "encoffining," in front of the family. In this way, family members have the opportunity to view the deceased clean and dressed in a white robe in the spirit of Buddhist departure. A former *nōkanshi* near Tohoku, Shinmon Aoki, describes this ritual in his book *Coffinman: The Journal of a Buddhist Mortician* (in Japanese, *Nokanfu nikki*), which became the basis of a popular Japanese film *Okuribito* (Departures).[23] His account depicts the intimate bond between living family members and the deceased as they pray for the deceased's journey to the land of the spirits, *kami.* It is therefore of paramount importance that the dead body be treated with the utmost respect and that it be cleaned in order to leave a favorable impression upon survivors. After the family viewing, there is the usual wake when the body in the coffin is viewed by neighbors, friends, and coworkers. After the wake, there is a Buddhist funeral service (often rather costly). After these services the body is then transported to the crematorium (*kasōba*).

All this points to the fact that deep-rooted cultural habits through customary rituals of death including the funeral and later ceremonies of prayer and remembrance are ways for survivors to come to terms with the departed's leave-taking. Japanese have always believed that there is a close link between those recently deceased and the living. Funeral ceremonies commemorate the loss of the deceased relative as family members pray for the soul's departure and smooth journey to the realm of spirits, *kami.* These prayers continue long after the cremation and burial of ashes. Buddhist altars, or *butsudan*, are still found in numerous houses throughout Japan, in a small corner of a room displaying the departed's photo, an offering of fruit, and lit incense for prayer. Japanese believe strongly that this link between the dead and the living continues and rests upon prayers we offer to the departed. In this case, time gently has its way of healing.

Soon after our Mom died, Dad had a small table in their Florida home upon which was a photo of Mom, flowers, and urn with her ashes. This was his special spot for Mom where he could visit each day and offer silent prayers. Now that he has recently passed away, we have in our home a special spot with a small urn containing some of Mom and Dad's ashes comingled next to their photo. Near that spot, we also have an urn bearing the ashes of Brooke's Mom and Dad, along with their photo.

There is a dark side to this, however. All this attention to the deceased can take away considerations regarding the living, the survivors. A survivor can feel so deeply alone and alienated that such ceremonies may carry little meaning. A throbbing issue when working with survivors lies in somehow offering professional counseling and providing psychosocial and emotional support for those who are particularly vulnerable and alone. March 11's all-encompassing destruction thwarted proper grieving from occurring, adding insult to injury for survivors who relied on these rituals as part of their grieving process. The sudden and horrible deaths that victims faced overwhelmed the senses and memories of survivors. We survivors often unintentionally visualize the last moments of our loved ones. And we pray deep in our hearts that their ending was quick and painless. Yet, for many survivors, prayers were no match for the unexpected horrors of 3.11 and the following weeks.

Perhaps the most terrible anguish for family survivors lies in not finding the deceased, either because the deceased has been so thoroughly crushed and buried within the mud, incinerated in widespread flames, lost somewhere, or dragged out to sea. There remain over 2,500 missing persons. Families are left to mourn without physical proof of death, without the presence of a body to whom they can offer last respects and prayers. Mothers of missing children particularly feel this anguish. Some mothers even acquired licenses to use large digging equipment for their search.[24] Years after 3.11, mothers can still sometimes be seen walking along the sea looking for evidence of their child. Their search occurs not only along the coast but on land, where they will scour the land looking for their child.

Put simply, without a body, closure for survivors remains extremely difficult. This is not a cultural feature. When my cousin Kevin Coyne took out his single mast sailboat on a freak winter day with his girlfriend and her pet dog, it was a sunny, cloudless day and the temperatures were in the 70s with only a slight wind. For some reason, their boat overturned and all three spilled into the icy ocean water off Newport, Rhode Island, where we all grew up. Within minutes hypothermia can set in. A neighbor spotted them in the water and, after around ten minutes, made it out to them in his tiny rowboat that could only carry two. Kevin insisted that his girlfriend and her dog get into the small boat as he hung on to the side of the capsized vessel. Both girl and dog were rowed ashore. All this took another 15 to 20 minutes. By the time the rescuer rowed out to Kevin, he was gone. He had lost all feeling and his grip gave way. His body was never found. Yet for years afterwards, his parents, Bud and Jean, would drive out to that same site at the Ocean Drive where Kevin was last seen, and look, wait, and hope. The sea takes, but rarely gives back.

Kasetsu Jūtaku and Our Human In-Between

Months after 3.11's terrible impact, temporary housing shelters, *kasetsu jūtaku*, were put together by the hundreds throughout the swamped regions. The term *kasetsu* combines *ka*, meaning "temporary," and *setsu*, or "facility." *Jūtaku* means "home for humans."[25] These are prefabricated units constructed of cheap material but enabling at least a minimum of decent housing. This is a major feature in the government's reconstruction program. These *kasetsu* are simple habitats with the basics—kitchen/living room, bedroom, and bathroom. Many of them are far inland and often out of the way which better protects them from future tsunamis. But dwellers face difficulties in accessing food, shops, mainstream community and community events, healthcare, and necessities. Very few survivors have cars. The sea or fire destroyed them. And there are very limited bus services. Now, with most of the train lines repaired, the ever-reliable Japan Railway's train service offers its steady public transit. *Kasetsu* dwellers, however, find it difficult to get to the train stations.

Who goes to these *kasetsu jūtaku*? With priority first given to families with small children, sick family members, and elderly, these units were later allocated by lottery to survivors who had been crammed into packed evacuation shelters. Survivors had lived in these evacuation shelters for months in close quarters lacking privacy. Here we have another version of "lifeboat ethics," conventionally applied in clinical healthcare settings having to fairly distribute scarce medical resources like antibiotics, organs, and ICU beds, when there are many patients in need of them. *Kasetsu jūtaku* are scarce resources, and allocating these will often be decisions of life and death, particularly for the many elderly squeezed away in evacuation centers' tight living quarters with all sorts of discomforts. For instance, elderly crammed in primary and elementary schools had to resort to using small toilets.

There was an egalitarian approach to allocating *kasetsu* through a lottery method, all given an equal chance. In effect, however, this split apart neighbors who had lived together and supported each other for generations. Close neighbors in Otsuchi, for instance, were separated from other surviving neighbors as they were placed into *kasetsu* at the town's outskirts, far from neighbors closer to the city center. Rural and coastal inhabitants found themselves estranged in the more urban settings, which, for many of them, are impersonal and consumer-driven. Closer to the damaged Daiichi nuclear complex, many who were relocated inland still had their homes somewhat intact, though heavily damaged. Yet they were without their own cars and therefore unable to attend to looking after, repairing, and cleaning their homes.

At first, *kasetsu* dwellers were told they would be in these temporary units for around two to three years. Most still live there, like Ryoichi Usuzawa. Ryoichi is one of the directors of the Tono Magokoro Net volunteer center where I worked in 2014. He is often seen in his donated car covered with blue, green, and pink flower stickers with messages of encouragement and support, most of which are written by children. Everyone knows Ryoichi. He and his family survived the tsunami, but his home and nearly all houses in his neighborhood in Otsuchi were swept away by the waves. Figure 1.1 shows a large barren area in Otsuchi over three years after the earthquake and tsunami.

Months after 3.11, suffering bouts of depression and intense feelings of isolation and guilt for surviving the disaster, Ryoichi decided not to allow himself to live in his cave of despair. As a way to help him recover from his depression, he now spends much of his time speaking to groups about the tragedy, his experiences, and the plight of survivors. For these survivors, many more years will pass before there is any semblance of rebuilding their towns and having their own homes. *Kasetsu* dwellers pay a slight fee to stay in these units. But the government gives them some basic items as well as up to 90,000 yen (approximately $900) per month. With many towns still flattened, restoration is many years away, probably decades. Restoration measures are tied up in bureaucratic hoops, politics, and power struggles over allocation resources. One plaguing question concerns land ownership. There continue to be lengthy, costly searches for identities of property and title owners. Bear in mind that the tsunami also wiped out town halls and official records regarding ownership, inheritance, and residents' identities. In addition to establishing rightful ownership of destroyed houses, a significant challenge lies in building affordable, quality homes with low-interest loans to help the victims who have lost all monies.[26] Some coastal towns may never see rebirth.

Figure 1.1. A barren Otsuchi three years later, March 2014

Can the region recover from this? The degree of recovery rests upon numerous factors. For instance, one key question lies in how long subsidies for *kasetsu* dwellers will last. Then there is the matter of jobs. With the region's fishing industry seriously fractured and taking the hardest hit, commercial fishing may take the longest time to recover. Local businesses, small shops, and industries are also seriously splintered. Many who are able to find work are employed in reconstruction and relief. But business sector prospects are appearing less bleak than the fishing industry. Employment has witnessed an upsurge. In 2012, for all three affected prefectures, unemployment was actually below the national average of 4.3 percent. One year later, in 2013, Miyagi's unemployment was 4.7 percent. Fukushima's is still below national average at 3.5 percent, and Iwate's is 3.4 percent.[27] Yet, when government stipends dry up for *kasetsu* and for recovery, what then? And add to this the monumental monies that are now being funneled into preparing for the 2020 summer Olympics. I describe this irony regarding recovery prospects and the 2020 Tokyo Olympics in my more recent June 26, 2014, *Albany Times Union* column, "Let the torch illuminate solidarity."[28]

It is nearly impossible to come up with the exact sum of casualties from the 3.11 triple-disaster. There are conflicting numbers, and data is incomplete. It is clearly the case, however, that fatalities, missing persons, those seriously injured, disaster-related deaths, and property destruction combine to make it Japan's worst catastrophe since the bombing of Hiroshima and Nagasaki. A report issued one year after the impact estimates that there were over 16,000 deaths. Over 90 percent are tsunami-related. Though the report does not specify deaths from exposure to the cold and hypothermia, these probably also factor into the fatality rate. The report states that the remaining 10 percent died from the earthquake with its resulting fires, landslides, and heart attacks. Nearly 3,000 remain missing. Those seriously injured number well over 50,000.[29] A more recent February 2014 report from the Japanese National Police Agency confirms 15,884 deaths and 2,602 missing.[30]

Of the three prefectures directly impacted by the disaster—from north to south, they are Miyagi, Iwate, and Fukushima—Miyagi suffered the most fatalities, with around 10,800. Approximately 5,800 people in Iwate died. And Fukushima experienced around 1,800 deaths. The coastal area extending one kilometer (.6 miles) from shore suffered serious damage, and the wave reached as far inland as ten kilometers (6.2 miles). That's six miles of extensive damage all along that major portion of the coast. Over 130,000 buildings were totally flattened (*zenkai*). 104,000 of these were from the tsunami. The earthquake caused the rest. Around 250,000 buildings were significantly destroyed (*hankai*), 140,000 from the tsunami, and 110,000 from the earthquake. Another 710,000 buildings were seriously and sufficiently damaged. Most of these, around 633,000, were the result of the earthquake.[31] Of the worst hit towns, Ishinomaki in Miyagi Prefecture heads the list, with over 2 percent of its population killed, around 3,735 deaths. Otsuchi has the highest percentage of residents killed, over 9 percent, with 1,286 deaths.[32] But these are numbers. The list goes on. So does the anguish.

NOTES

1. Lucy Birmingham and David McNeil, *Strong in the Rain: Surviving Japan's Earthquake, Tsunami, and Fukushima Nuclear Disaster* (New York: Palgrave Macmillan, 2012), 128–29. The authors, journalists based in Tokyo, offer an excellent account of the disaster.
2. Birmingham and McNeil, *Strong in the Rain*, 6.
3. Tom Gill, Brigitte Steger, and David H. Slater, "The 3.11 Disasters," in *Japan Copes with Calamity: Ethnographies of the Earthquake, Tsunami and Nuclear Disasters of March 2011*, 2nd ed., edited by Tom Gill, Brigitte Steger, and David H. Slater (Bern, Switzerland: Peter Lang, 2013, 2014), 3.
4. Saint Joseph Edition of *The New American Bible* (New York: Catholic Book Publishing Co., 1970), 21. The Bible is silent as to the name of Lot's wife, though some Jewish traditions later referred to her as "Idit" or "Edith."
5. Gill, Steger and Slater, "3.11 Disasters," 4.
6. Birmingham and McNeil, *Strong in the Rain*, 71.
7. Daiichi is often referred to as Plant 1, and Daini as Plant 2. *Ichi* and *ni* mean "one" and "two" respectively.
8. Ranjay Gulati, Charles Casto, and Charlotte Krontiris, "How the Other Fukushima Plant Survived," *Harvard Business Review* (July 2014), accessed Dec. 17, 2014, https://hbr.org/2014/07/how-the-other-fukushima-plant-survived.
9. Gregory Clancey, *Earthquake Nation: The Cultural Politics of Japanese Seismicity, 1868-1930* (Berkeley, CA: University of California Press, 2006), 104.
10. National Catalogue of the Active Volcanoes in Japan, July 2013, accessed Dec. 16, 2014. http://www.data.jma.go.jp/svd/vois/data/tokyo/STOCK/souran_eng/menu.htm. Also accessed Dec. 16, 2014. https://gbank.gsj.jp/volcano-QV/VOL_JP/EN/active_v.htm.
11. "Japan Dotted with Volcanoes that Could Erupt at Any Time," Nikkei Asian Review, accessed Dec. 16, 2014. http://asia.nikkei.com/Tech-Science/Science/Japan-dotted-with-volcanoes-that-could-erupt-at-any-time. The Coordinating Committee for the Prediction of Volcanic Eruption, the group of experts who assess volcanic activity, has changed over time in determining what constitutes "active."
12. Bruce Parker, *The Power of the Sea: Tsunamis, Storm Surges, Rogue Waves, and Our Quest to Predict Disasters* (New York: Palgrave Macmillan, 2010, 2012), 151–52.
13. Parker, *Power of the Sea*, 151.
14. Actually, Mexico's warning system went online in 1991, well before that of Japan's which was launched in 2007. Just before a magnitude 7.4 rattled Mexico City in March 2012, Mexico's warning system proved effective in enabling safe evacuation.
15. "Comparing Earthquakes, Explained," Robert Coontz, accessed Dec. 14, 2014, http://news.sciencemag.org/2011/03/comparing-earthquakes-explained.
16. "Magnitude/Intensity Comparison," U.S. Geological Survey, accessed Dec. 16, 2014, http://earthquake.usgs.gov/learn/topics/mag_vs_int.php.
17. Parker, *Power of the* Sea, 151.
18. Frances Ashcroft, *Life at the Extremes* (London: Flamingo, HarperCollinsPublishers, 2001), 149.
19. One passenger who survived the crash, Arland Williams, drowned after passing lifelines to others but not keeping one for himself. The cold water soon overtook him. Later on, the bridge was renamed the Arland D. Williams Jr. Memorial Bridge.
20. Ashcroft, *Life at the Extremes*, 152.
21. Ashcroft, *Life at the Extremes*, 161.

22. Michael C. Brannigan, "A Chronicle of Organ Transplant Progress in Japan," in *Transplant International* (Vol. 5, 1992), 180–86. Oddly, the Japanese still do not officially sanction brain death. The only instance in which brain death is permitted is in the case of a heart donor consenting to be declared brain dead so that his heart can be officially removed.

23. Shinmon Aoki, *The Journal of a Buddhist Mortician: Coffinman* (orig. *Nokanfu nikki*), trans. Wayne S. Yokoyama (Anaheim, CA: Buddhist Education Center, 2002).

24. Birmingham and McNeil, *Strong in the Rain*, 152.

25. Kunio Aoki and I discussed the literal meanings of *kasetsu jūtaku*, March 21, 2014.

26. Gill, Steger, and Slater, "3.11 Disasters," 9.

27. Gill, Steger, and Slater, "3.11 Disasters," 11. Compare this with Rhode Island's high unemployment rate of over 8 percent in 2014.

28. See Michael C. Brannigan, http://www.timesunion.com/opinion/article/Michael-Brannigan-Let-torch-illuminate-solidarity-5582346.php.

29. "Japan—366 Days after the Quake . . . 19,000 lives lost, 1.2 million buildings damaged, $574 billion," Earthquake-Report.Com, accessed Dec. 15, 2014, http://earthquake-report.com/2012/03/10/japan-366-days-after-the-quake-19000-lives-lost-1-2-million-buildings-damaged-574-billion/.

30. "Damage Situation and Police Countermeasures," National Police Agency of Japan, accessed Dec. 15, 2014, http://www.npa.go.jp/archive/keibi/biki/higaijokyo_e.pdf.

31. Gill, Steger and Slater, "3.11 Disasters," 6.

32. "Damage Situation and Police Countermeasures."

Chapter Two

Precariously In-Between: Desperation and Moral Grit

Japan's unfolding anguish did not start with the 3.11 calamity.

"3 a.m. This human being hasn't eaten in 10 days but is still alive . . . I want to eat rice. I want to eat a rice ball." These were the last words of a fifty-two-year-old man who starved to death in his home in Kitakyūshū, located in Fukuoka Prefecture on the southern island of Kyushu where I was born. His diary of his last days tells it all, an agonizing ordeal of simply trying to live without food. Unemployed, his welfare benefits cut off, and his monies dried up, in the end he didn't even have enough money to buy *onigiri*, a rice ball. His partly desiccated corpse was found about one month later. This was in summer 2007. In the same city, a similar scenario had played out in each of the past two years in which a solitary man, unemployed, completely alone, hungry, and without money and benefits wasted away in starvation, only to have his mummified corpse discovered months later.[1]

In addition to the countless homeless in Japan who live and die in parks and in dark city corners, more people die alone, many of them "homeless" at home, *hikikomori*, suffering pain from the day-to-day grind. In Tokyo alone, there are around ten reported suicides a day. 31 percent of Japanese live alone. Of these, over 23 percent are over sixty-five years of age.[2] This is well before 3.11. More recent figures confirm that Japan still has one of the world's highest suicide rates. In 2013, there were 27,283 reported suicides. This means someone takes his or her life every twenty minutes. These reports also indicate that men commit the highest number of suicides, nearly 69 percent, and the highest incidence is among Japanese in their sixties, 17 percent. The leading factors in suicide include health issues at the top, followed by financial, family, and work problems.[3]

For decades leading up to the triple-disaster, a festering sense of solitariness and bleak desperation, what cultural anthropologist Anne Allison calls "precarity," has been slowly creeping into Japanese society.[4] According to Allison, this "precarity" has been overshadowing Japanese society, despite the country's "economic miracle" after it nobly and orderly "embraced defeat" in the Pacific War.[5] In past decades, people are becoming increasingly socially disconnected, isolated, and withdrawn. I saw signs of this years ago when I lived in Westchester County, New York. Many Japanese moved there when companies sent husbands to work temporarily in New York City. They and their families lived a life in transit. Though many relished their newfound freedoms and larger homes, many were also disconnected from their native land. In addition to my college teaching, I was offering English-speaking lessons to families of some of these transients who commuted daily to Manhattan. For months, a young Japanese boy, twelve years old, whose family lived in Scarsdale, never came out of his room except to use the bathroom. He missed his classes, and only ate his food when it was placed on a tray outside the door to his room. His mother felt helpless and did not know what to do. She sought my help in urging him to leave his own room. This became for me a daily challenge. In place of my weekly English lessons, I would go to the house as often as I could to speak with him and coax him out of his room. Eventually, on a few occasions he did step out, and we spoke only briefly. I urged that the parents see a professional counselor. They were about to have a meeting with a therapist when the father's company sent him and his family back to Japan. They never gave me their address in Japan. I often wonder whatever became of the son.

Consider employment. Work, in Marxist strokes, defines how we become. It gets to the core of the relationship between living and labor. For Japanese, especially men who are hard-wired to work, their lives are absorbed in work. "I work, therefore I live." Work and its all-consuming routine eats into their lives. After its remarkable post-war recovery, Japanese have lived under the shadow of a secure contingency. In a land where work eventually absorbs the soul, as in the case of the *sarariman*, as my mother used to tell us when we were young, "one lives to work" more than works to live. (This was one reason she confessed to us that she would have never married a Japanese man.) But at least the context was some measure of permanence in that workers gained lifetime employment, security for self and family. But the tides have shifted. The economic bubble has burst, and there is no longer any guarantee of working tomorrow. Japan no longer has its practice of life-time employment. One-third of Japanese are "irregularly" employed, that is, with part-time contracts and without benefits of security. Of these, 77 percent of the salaries are below the poverty level.[6] And poverty, *hinkon*, in Japan affects 20 percent of the country's elderly. Reserves are depleted. This pre-3.11 desperation is evident in the high stakes gamble of towns surrendering to nuclear politics and permitting nuclear sites in their backyards in exchange for the carrot of nuclear industry jobs, at highly vulnerable sites, perversely restoring identity through work. An increasingly all-powerful nuclear industry exploits an aging population and dying economy. This is a lethal bargain: exchanging identity/work for the risk of contamination, especially ironic considering the radiation poison rained upon them (black rain) at Hiroshima and Nagasaki. Desperation runs deep. Today, now more desperate than ever, many men, even though unskilled, sign up to work at decontaminating the Daiichi nuclear plant. Wages are low, but it puts food on the table.

Although the country still has the world's third largest economy, it now owns the second highest rate of poverty among Organization for Economic Cooperation and Development (OECD) nations. Japan's 15.3 percent poverty rate is second to the U.S.'s 17.1 percent.[7] For many Japanese, with the burst of their economic bubble and growing sense of insecurity and instability, this material insecurity has led to an existential floundering, as depicted in Kobo Abe's (1924–1993) popular novels with titles like *Woman in the Dunes*, *Man in a Box*, and *The Face of Another*. Abe often depicts his protagonists in surreal circumstances to further underscore the absurdity of the human condition. (This is distinct from Albert Camus's *l'absurde*, which referred to the absurdity of existence itself, more so than the human condition.) His Box Man (*Hako Otoko*) decides to withdraw from his work and wears a cardboard box over his head through which he views life in the frenetic city of Tokyo. In his *Woman in the Dunes* (*Suna no Onna*), the teacher and entomologist Niki Jumpei lives with a woman whose house is in a sandpit, in constant danger of being buried in sand which Niki shovels incessantly to prevent the massive dunes from crushing the house and community. Abe's perennial themes center around alienation and absence of personal identity. Often, this leads to struggles in human interaction and communication. Reflecting the prevailing social mood, especially during the literary period of the Theater of the Absurd in the 1960s, represented by Samuel Beckett, Eugene Ionesco, and Harold Pinter, it comes as no surprise that Abe's novels were bestsellers.

Matters of identity are intimately related to one's place, environment, space, and relationship to others. All this starkly reflects our inescapable in-betweenness, our being situated, helplessly, within climates, environments, landscapes, seascapes, soundscapes, and human-scapes. Along these lines, for Japanese, 3.11's triple-disaster offers no new lessons, for this became increasingly evident prior to 3.11. Well before the events and aftermath of March 11, 2011, Japan was losing its tight grip on traditional cultural values, including its imbeddedness in relationality. All this shatters the popular stereotype we outsiders may have of a collective, socially bonded and cohesive Japan. And these "lonely deaths," *kodokushi*, continue to escalate in Japan.

DISLOCATION'S PAIN AND SUFFERING

In what activist writer Amamiya Karin terms *ikuzurasa*, the "pain of everyday life," contingency remains the dominant card.[8] Displacement or "dislocation," the literally meaning of the Buddhist *dukkha* which is usually translated as "suffering," is all the more Japan's reality. The 3.11 tragedy was the latest wave of dislocation, imprinting further the stubborn truth that all is fragile and tenuous. When life equals work, the death of work spells the death of life. Socially and economically, Japan was conducting an ongoing dress rehearsal for 3.11. March 11, 2011, was the capstone event. Though its scope remains unimaginable, it lives on for countless Japanese who endure lives of *ikuzurasa*.

Sixty-four-year-old Hisashi Tarukawa, a native of Fukushima, grew cabbages on his farm in Sukugawa, seventy kilometers, forty-three miles, away from the Fukushima Daiichi nuclear plant. Weeks after the tsunami crippled the plant, the government warned against buying any cabbage from Fukushima. His cabbages survived the tsunami, but he was banned from selling them due to risks of high radiation levels. Tarukawa's livelihood was in growing healthy cabbages, avoiding chemical fertilizers. After the government's announcement, the farmer knew that for him, all was lost. The earthquake shattered his home. And the threat of radiation killed any hope for rebuilding a future in farming, which was his passion, his life. Two weeks after the impact, on March 25, out of sheer desperation, Tarukawa hung himself.[9]

On June 11, a dairy farmer in Soma, Fukushima Prefecture, around 45 kilometers (nearly 28 miles) from the nuclear plant, after realizing that he could no longer sell milk, hung himself in his cattle shed. There, on the wall of the shed, he scribbled his final words: "If only there wasn't a nuclear power plant . . . I have lost the energy to carry on working."[10]

As of this writing in 2014, more than three years after March 11, 2011, Tohoku's suicide rate intensifies as people live daily on the edge. For thousands, the well is dried. They've lost family, home, livelihood, and neighbors. Signs of this loss were evident with ubiquitous homemade posters not only of missing family members and neighbors, but missing pets and farm animals. Already having one of the world's highest suicide rates (around 30,000 annually), Japan witnessed a dramatic increase soon after the 3.11 tragedy. This was particularly the case in Miyagi Prefecture. There, suicides increased nearly 40 percent according to some accounts.[11] Many of those who survived not only lost homes, but families and livelihoods. Rice farming had always been bountiful near the Sanriku coastline, producing rice for much of the region and other areas. When radiation from the Daiichi nuclear power plant spewed forth into the atmosphere and spilled into the soil, most nearby farms were completely ruined. And without their farms, their means of living were ruined. The same was the case for farms with livestock and crops. Suicide victims included farmers, elderly, and even young people who had lost their families. I met the innkeeper Akiko Iwasaki in Kamaishi in January 2012 at the reopening of her *ryokan* (traditional Japanese inn). (We will meet her in chapter 4.) She was with her daughter and infant granddaughter. Akiko and her daughter spoke to me of her daughter's best friend and former high school classmate who, though she survived the tsunami, had been missing for the past few months. She had lost both of her parents, her siblings, and many friends from school. She was last spotted walking alone along the scenic and lonely span of beach just outside of town, near the *ryokan*. She has never been found.

Managing Grief

Surely, the magnitude of survivors' loss is overwhelming. The disaster wiped away their loved ones, families, friends, and neighbors. Many are left to grieve alone, like the high-school girl. And grieving alone is a large piece of the bigger problem, that of managing one's grief. Here in the United States, in our more heavily individualized society, for every death there are around seven survivors at risk, persons especially close to the deceased. They are at risk in terms of how they manage to mourn, process, and come to terms with the deceased's death. How they do so determines whether their grief is healthy or unhealthy. For this reason, grief counseling is critically important for us.

In a society like Japan that is much more collectively oriented, for each person who dies, the circle of persons at risk extends outward much further. And these persons at risk are in jeopardy, depending on how they are able to come to terms with the death. And here lies a dark side to long-standing cultural habits. As a culture, grief counseling is neither strongly urged nor supported throughout much of Japan. This is particularly so in the more rural areas like Tohoku. There, men especially resist any form of therapy or counseling. This generates all sorts of difficulties for survivors. And most suicides are committed by men. For many Japanese, their sense of collective identity is such that individuals will often suppress their grief in order not to burden others with their sorrow. As bad as one's misfortune is, he can reason to himself that others are worse off. This is especially the case in Tohoku, which historically has endured more than its fair share of tragedies, famines, and poverty. Yet connectedness with each other and with the community has enabled the people of Tohoku to respond with resilience and solidarity. However, while this suppression of personal grief and sorrow for the greater good may enable a temporary capacity for some measure of collective and community support, it also takes its toll.

Norway is another society that, like Japan, is collectively oriented. It is much more open, however, to counseling and group support. On two separate occasions I met with Anne Cecilie Fossum, Oslo's Red Cross director who played a key role in response to the July 22, 2011, Oslo explosions and Utøya camp massacre. She also led in facilitating support for the victims' families. At our first meeting, she stressed the powerful influence of Norway's emphasis on solidarity, particularly in its growing climate of multiculturalism. One of the young victims at the camp was Iraqi. His father attended the trials. At one session, when confessed gunman Anders Breivik took the stand, the father became so upset he threw his shoe at Breivik. All in the courtroom were shocked. This was very un-Norwegian. After a pause, however, everyone softly applauded to show their empathy, understanding, and support.[12] In my second meeting with Anne Cecilie, in May 2014, she noted that three years after the tragedy there was an evident increase in issues related to grieving. At first, the disaster was so publically exposed, personal grief was put on hold. However, this suppression on the personal level has generated longer-term concerns. As of 2014, according to Bergen's Crisis Psychology Center report, over one-third of bereaved parents have still not returned to work and over 40 percent of victims' siblings experienced a drop in their school grades.[13]

What compounds the management of grief all the more in Japan, a culture with a long-standing emphasis on avoiding shame, is its enduring historical habit of minimizing issues of mental health. For centuries, problems related to mental health were often considered taboo and unspeakable. Renowned novelist Kenzaburo Oë writes about this in his novel *A Personal Matter*, a tale of a twenty-seven-year-old teacher nicknamed "Bird" whose wife gives birth to an abnormal baby. The abnormal baby is believed to be the result of the atomic bombing's ensuing radiation poisoning among survivors. The thought of being burdened with a mentally disabled child horrifies him to the extent he even ponders suicide. Caring for the baby will also be an obstacle to his lifelong dream of going to Africa. At first, Bird is opposed to an operation that would save the baby of its alleged brain tumor (later discovered as benign). So too, medical doctors display ridicule towards the baby. Here, medical professionals reflect society's repugnance towards mental disability. Gradually, however, he awakens to his duty as a father to save the baby by insisting on the operation and then caring for the child.[14]

Japanese filmmaker Kazuhiro Soda recently brought the plight of the mentally ill to the public with his 2008 documentary *Seishin* (Mental).[15] This was sparked by his own lived experience with a mental condition, "burnout," and enduring public and professional stigma. The film depicts patients in

Chorale Okayama, a small mental health outpatient clinic in Okayama Prefecture. The clinic, compared to others in the country, has a far more progressive approach towards treating persons with mental disorders. By unveiling the patients' personal stories and anguish, the film was instrumental in raising efforts to increase public awareness and offer more professional counseling services. At the Dubai and Pusan film festivals, *Seishin* won the Best Documentary Award.

Clearly, public perceptions regarding mental disorders are influenced by the public's understanding of mental illness. The fate of children who have lost one or both parents from 3.11 is a case in point. Many children lost parents to the tsunami. Over two hundred lost both parents. Table 2.1 below shows the current count.

Many school children lost their mothers to the waves when their mothers drove to the schools immediately after the quake's shuddering. *En route* to the schools, mothers were crushed by the waves as they frantically tried to bring their children home (see table 2.1). In Rikuzentakata alone, where I spent much of my time volunteering, forty-one children were orphaned. Adoption by foreign nationals is unheard of in Japan, as orphaned children will usually be taken care of by living relatives. Those without any existing relatives are cared for by foster care homes, often sponsored by Christian churches, like the Catholic La Salle Home in Sendai.[16]

Makiko Okuyama heads the Department of Psychosocial Medicine at the National Center for Child Health and Development in Tokyo. I spoke with her along with my friend and chief of their Division of Health Policy and Bioethics, Naoko Kakee. I had known Naoko for years and we share a strong interest in bioethics. She and I had often discussed the dynamics of how U.S. mainstream bioethics has developed in Japan. Naoko introduced me to *matcha* tea, the tea powder used in the traditional tea ceremony, *Cha no Yu*. Naoko graciously introduced me to Makiko. Makiko, a renowned and well-respected scholar, has spent much time with her research team after 3.11 studying the effects of trauma particularly upon surviving children. She and her team worked in stricken areas with the children, closely observing their

Table 2.1. Parental Loss by Prefecture

Prefecture	Orphans	Children who have lost one parent
Iwate	94	575
Miyagi	135	1047
Fukushima	21	160
Total	250	1772

Source: Makiko Okuyama, National Center for Child Health, Tokyo, March 24, 2014.

behavior, their adaptability to their new situations, for example in *kasetsu*, and looking for signs of post-traumatic stress. She admitted to me, however, that this was no easy task since cultural behavior and beliefs, particularly in that part of Japan, Tohoku, tend to downplay mental disorder after severe trauma. Instead, cultural beliefs tend to underscore the importance of remaining stoic even after extreme duress. This applies to children as well. Nonetheless, Dr. Okuyama and her team noticed clear symptoms of post-traumatic stress disorder (PTSD) among many of the surviving children. Her team was able to work with many of these children and assist them in their grieving process. Thanks to them, many children received sound counseling and are now better able to cope with their memories of those terrible days. At the same time, Makiko emphasized how grief counseling is still all the more urgent when working with children, adolescent, and adult, particularly elderly, survivors.

WATSUJI TETSURO—OUR INESCAPABLE IN-BETWEEN

Japan's precarity starkly illustrates our inescapable in-between. It reminds us of our unavoidable imbeddedness in links—global economy, capitalism, communication, and so forth—that lie beyond our immediate control. And yet, because of our imbeddedness, affect us in nearly every corner of our individual and collective lives—social, medical, technological, educational, and material. Allison's notion of precarity resembles eminent sociologist Zygmunt Bauman's "liquid world," in which the traditional forces of stability such as religion, national sovereignty, and family have dissolved under the tsunami of global capitalism, whose epicenter foments from Western, more often U.S., rumblings.[17] Whether we call all this "precarity," "liquid world," or Anthony Giddens's "runaway world," it all amounts to the same thing—all is replaceable, impermanent, ever-poised for disconnection.[18] This differs from our usual stereotype and impressions of a relational, connected Japan. Precarity and liquidity, our pronounced insecurity, represents the Buddhist idea of *dukkha*, our inherent dislocatedness, which in turn represents the shadowy side of our living "in-between," what Japan's eminent philosopher Watsuji Tetsuro calls *aida*, or *aidagara*. And suicide illustrates the dark extreme of our indelible *aidagara*, in-betweenness in which we live unavoidably in-between past, present, and future. What feels like the utter annihilation of all that represents living—family ties, closeness, and work—can provoke what Jonathan Lear describes as the "loss of subjectivity," a loss of particularity that disrupts one's vision of the future, resulting in meaninglessness and despair.[19] This challenges us to seriously consider how we go about managing this truth when we face trauma and loss. So far, we've only alluded to this "in-between." Let us now unpack this notion further. There's no better way than to examine its force in the writings of Watsuji Tetsuro.

Discovering Ourselves through Physical Climate

Watsuji Tetsuro (1889–1960) is Japan's most prominent philosopher of ethics. His ethical theory centers uniquely around the significance of climate. Just wrap our heads around that—a philosopher writing on the importance of physical climate. This is natural today for philosophers engaged in ecology and the environment. But not so in Watsuji's time. Philosophers then and now tend to face that occupational hazard of taking one's ideas too seriously, living in them, and bestowing mental constructs with their own reality, rather than stepping outside of them into—weather and climate.

As a young scholar, Watsuji was first drawn to Western philosophy's emphasis upon individualism, independence, autonomy, and freedom. This was a refreshing departure from what he felt was an oppressive conformity in Japanese society. The writings of Kant, Nietzsche, Kierkegaard, Heidegger, and others excited him to the possibilities in individual human freedom and expression. However, as much as he admired Heidegger, he became deeply critical of what he believed to be Heidegger's overemphasis upon temporality as a condition for human authenticity. For Watsuji, the idea that we are essentially "beings in time" was one-sided. For him, Heidegger undermined the significance of our spatiality, that we are also beings in space, in an environment, in climate. And his writings set out to underscore our situatedness in place and particularly among other persons. Here, Sōseki Natsume's thought, especially his stress on our fundamental betweenness among others, sets the stage for his revisiting Japanese teachings on community and the importance of finding a proper balance between the individual and the individual's innate relationality with others.[20] Watsuji thus rejects Western style individualism—the "I" as independent and separate—in favor of our deep-rooted relationality and interdependency.[21]

Watsuji unpacks the idea of our inherent relationality in his pioneering *Fudo*, translated as "Climate and Culture." Here, he focuses on the all-important impact and influence of climate—geophysical, weather, etc,—on how we see ourselves, our world, and our relation to that world.

> We have discovered ourselves in climate, and in this self-apprehension we are directed to our free creation. Further, it is not only we ourselves who today cooperate to defend ourselves or work against the cold, the heat, the storm or the flood. We possess an inheritance of self-apprehension accumulated over the years since the time of our ancestors . . . We must accept that it is only through the interpretation of historical and climatic phenomena that we can show that these phenomena are the expression of man's conscious being, that

> climate is the *organ of such self-objectivization and self-discovery and that the climatic character is the character of subjective human existence.* (author emphasis)[22]

Watsuji examines cultures in terms of their climate, embracing weather, landscape, and seasonal features, all combining to fashion distinct worldviews of nature, self, other, and their symbiotic entanglement. *Fudo* essentially addresses the relationship between culture, cultural worldviews, and natural phenomena of climate.

In this work, he does not distinguish between climate and weather, as meteorologists and climatologists today do. For us, weather is about what will happen today, tomorrow, weeks ahead and into the season. Meteorologists forecast the weather. Climate has to do with weather patterns over a course of time—months, years, and decades ahead. Climatologists forecast what weather patterns indicate for the future. Their timescales differ. For Watsuji, it was evident that climate manifests itself in weather, day-to-day, and in patterns over a period. There is an undeniable pull which climate in terms of weather has on our state of mind and psyche. Though not unique to Japan, it is often the case that Japanese refer to inner emotional states using weather metaphors. Atmospheric terms like *uttōshii* (gloomy) and *sugasugashii* (invigorating) also convey inner states. Sunshine has a way of affecting the temperament of inhabitants, their inner temperature. Aside from the stock of vitamin D that a sunny day offers, it influences moods ranging from uplifting and cheerful to dull and lethargic when there is an excess of sunshine accompanied by high humidity. Commuter traffic can become pleasant and civil as well as near-insane with inflamed tempers. I spent years in Belgium for my graduate studies. Belgium is not known for sunny weather. It rained often with its fair share of gray, overcast skies. Actually, it offered for me a perfect climate for studying, or at least harboring inside a quiet room with my books, or in a lively tavern, more often the latter. Both offered solace from the weather. That one day when the sun shone, however, brought Belgians outside on the sidewalks, faces tilted towards the sun, basking in its warmth and making a temporary difference in their outlooks. No doubt about it: a change in the weather, however slight, can bring about a change in our inner climate, and this, overall, affects how we see the world.

Watsuji's fascination with climate comes as no surprise. He was intensely interested on examining our natural interconnectedness, what Buddhists refer to as *pratityasamutpada*. Though complex, *pratityasamutpada* in its bare skeleton means that all things without exception rise and fall in a symbiotic, mutually permeating relationship of cause and effect. Nothing exists in isolation, even those elderly during Chicago's terrible 1995 heat wave trapped in their city apartments. We'll read more of that below. All of life participates in this rhythmic, mysterious dance in which every step defines and is defined by all other steps.

Perceiving Risk

With the increase in extreme weather and more recent "acts of God," by necessity we've become more attentive to weather and climate. Weather definitely jumps to our screens. We wonder what the weather will be like today, tomorrow, and for the rest of our week. Will it rain, snow, be sunny, cold, hot? Weather concerns itself with the immediate. And the immediate matters, well, immediately to us. Climate, on the other hand, at least in meteorological perspective, pertains to the long-term, habitual play and pattern of weather. It concerns the bigger picture, the view from over 36,000 feet.

But we humans have difficulty viewing things from a long-term perspective. Take global warming. Experts have gleaned more than enough evidence to point to global warming as a culprit behind the extreme weather we've encountered in recent decades. If so, then the real perpetrator lies in human hubris as we continue to contribute to the build-up of greenhouse gas emissions in the atmosphere. Despite this, in a 2009 survey conducted by the Pew Research Center for the People and the Press, in a list of national priorities that include economy, jobs, education, healthcare, and global warming, those surveyed rated global warming at the bottom of concerns.[23] Why is it that global warming is situated last on the scale of priorities for Americans? Is it too apocalyptic, too scary?[24] Too abstract, though effects seem regularly concrete? The more abstract the concern, the more distanced we feel from it. Teaching theoretical ethics via mainstream Aristotle, Aquinas, deontology, utilitarianism, and social justice is clearly a challenge in this respect. Students begin to "get the feel" of moral concerns through the hook of concrete, contemporary cases and application. Medical students are a notorious case in point. As for philosophy, which we need today more than ever, its inherent abstraction lends itself to an early demise in the classroom. Is our response to global warming more of a denial on our part? Are we unwilling to face our environmental responsibilities and take them to heart? Or is the issue deeper in that we have an innate inability and/or unwillingness to think long-term? Are we pulled by the lure of immediacy? Like the lure that seduces some of us into buying beach front property in hurricane zones. This is dangerous, for it appears to reflect an increasing societal and cultural narcissism—all that matters is me, my own interests, and my own time zone. Thinking global warming demands extracting ourselves from this mind-set.

It seems to me that, of these many reasons, our inability as well as unwillingness to conceptualize in long-term ways stands out as the strongest. And this is further supported by the incessant power of our emotions in our perception of risks. That is, when we process risks, our feelings trump reason. Emotions trump reason when it comes to processing risks, their levels of seriousness, lethality, and likelihood, despite the evidence. Consider the current outbreak of Ebola, the worst of its kind in the disease's history. I write about this in my *Albany Times Union* November 26, 2014 ethics column, "Guarding against a deadly outbreak of unfounded fear."[25] What we think of as dangers "far down the road" or "over there" present less of a concern than those dangers facing us right around the corner. Dangers right around the bend and those in front of us more easily rattle our emotional equilibrium. And, of course, media's presentation of these dangers in 24/7 non-stop flow of instant information, images, and misinformation adds to our angst. Immediacy entices us far more than what we believe to be remote.

Moreover, we remain creatures of habit. Our day-to-day routine grind generates the ground of our lived being. We cultivate for ourselves the customary, an unfolding reciprocity between our habit and inhabiting, what we are accustomed to do and what we will do. Yet there is a certain irony in this. We have enormous difficulty in thinking long-range or long-term. However, we actually cultivate for ourselves this long-term through our everyday habits. Each day we build the groundwork for the future. We imprint our marks upon a long-range canvas, which, far from empty, encompasses each action we take right here and now. We can only act in the present, as that is all there is. We only see the present. Yet, how we inhabit this present will inevitably unfold in that distant not-yet which we can neither see nor conceptualize. Our habits are therefore another major determinant, perhaps the biggest, of risk-perception. Our past experience with a disaster can move us in wise directions—being well-prepared beforehand with sufficient supplies, moving out of Dodge, and so forth—or it can short-circuit our ability to factor in the worst case scenario simply because we've ridden out the storm before, and we can do it again. One powerful lesson from Hurricane Katrina is that the worst case scenario can and does happen. Black swans exist.[26] This latter lack of imagination partly explains the Daiichi power plant management's and Tokyo Electric Power Company's (TEPCO) lack of preparedness for a tsunami of its size. In a worst case scenario, the prospect of situating back-up generators in buildings' lower levels defies reason.

Encountering Ourselves through Social Climate

Fudo was a big first step in the development of Watsuji's understanding of ethics. His key work in ethics comes later with his *Rinrigaku* (Study of Ethics), subtitled *The Significance of Ethics as the Study of Human Being*, a landmark work in Japanese ethics.[27] An exceptional work as well as outstanding analysis of the relationship between human identity and morality, this work earned him recognition as a world-class philosopher.

The work is complex. Nonetheless, here is a concise summary that should at least enable us to view Watsuji's relevance to our study. First, Watsuji reminds us that the Japanese term for "human being," or "person," *ningen*, combines two distinct characters. The first character, *nin*, means "individual" or "individual human being." When used by itself, the character is called *hito*, still meaning "individual human being." The second character, *gen*, means "between" or "space." When that same character is used alone, it means *aida*, or *aidagara*, meaning "in-between."[28] So, "human being" or "human person," *ningen*, means that the human is both individual and "in-between," in relationship, situated among others. And this situatedness is inherent, unavoidable.

人間— *ningen*, human being or human person

人— *nin*, also *hito*, individual

間— *gen*, also *aida or aidagara*, in-between, space

We each have a dual nature. While we are each unique and individual, we are also innately related to all else. We do not exist as independent, separate beings. We exist *interdependently*. It is this quality of *aida* that underscores our relatedness and interdependency, our inescapable in-between. And this in-between is both spatial and temporal. We can thereby think of *ningen* as *ningen sonzai*, "human being in time and space." *Sonzai* combines two characters, *son* meaning "time," and *zai* meaning "space."

This constitutes the fundamental paradox of who we are—both individual and related. Each pole is in necessary tension with the other. We are each distinct from each other. Yet our distinction assumes a natural unity. Distinction can only exist within the context of oneness. This echoes earlier philosopher Nishida Kītaro's (1870–1945) pivotal teaching regarding our "identity of self-contradiction."[29] The distinction between self and other, which I refer to as our "original fault line," is a paradox since it assumes common ground in order for there to be distinctness.[30] When we apply this to homeless survivors of Japan's 3.11, this means that they surely feel their separation, their isolation, and it defines their everyday living. Yet, for many of these survivors, their feeling of isolation, of distinctness, is so strong they overlook their innate ties to others. Only concerted reminders of their connectedness can help them restore the whole picture of our paradox.

Our in-between is therefore the condition for the possibility of personal encounter. In like manner, Emmanuel Levinas (1906–1995) describes the moral significance of the Other's face, *le visage,* the face representing our natural in-between, and thus an invitation to an encounter, indeed a moral encounter.

> The epiphany of the face is visitation . . . The Other who manifests himself in the face pierces, in a way, his own plastic essence, like a being who opens a window where, nevertheless, his face is already traced. His presence consists in divesting himself of the form that nevertheless already manifested him . . . This is what we describe with the words: the face speaks. The manifestation of the face is the first discourse . . . the presence of the face signifies an irrefutable order—a commandment—that arrests the availability of consciousness.[31]

Philosopher Robert E. Carter beautifully captures this possibility for encounter:

> It is that precious sense of being available, being open to the possibilities of something worthwhile in the making, that is so characteristically Japanese. There will never be another meeting exactly like this one, in this time, in this place, with this person, with these personal histories: cherish it, like the cherry blossoms which may be gone with the next breeze, the next storm. Make of it a haiku poem, a snapshot of the moment, captured for all time, and then let it go in order to be ready for the next moment.[32]

In *Rinrigaku*, Watsuji makes it clear that climate is also interpersonal, social. This was a theme in his *Fudo* as well, but he elaborates much more extensively in *Rinrigaku*. Just as we live in-between nature and its phenomena, expressed in weather, atmosphere, geography, and so forth. and are thus subject to nature's never-ending manifestations, we also live in-between other inhabitants in relation to nature. We inhabit an unavoidable social climate that consists of other living beings in relation to each other. We live in a web of social geography, imbedded in relationality to, with, and for others. On all accounts, this too is inescapable.

We first discover this with the family in which we unintentionally find ourselves, without our consent. A striking example of this is Yukiko Fujimoto. Yukiko is one of countless adults displaced by the 3.11 tragedy, split apart from their families. She is split between her daughter who lives in Tokyo and her mother who lives in Sendai. I met Yukiko at the Kunuma train station where she took the two hour ride from Tokyo to meet me and Kunio in 2014 after my volunteer stint. We were on our way to the famous town of Nikkō, where Yukiko did the interior designing of the town's popular *ryokan*, Tsurukame Daikichi, in the heart of Nikkō.[33]

Yukiko is an elegant and striking grandmother of four. We later discovered she actually has indirect lineage to the emperor. Her daughter's husband's grandmother is sister to Emperor Akihito. The 3.11 tremblor completely destroyed Yukiko's home in Sendai. She managed to escape unharmed, but she lost many friends and neighbors to the ensuing tsunami. She is torn between her mother and daughter. Her mother still insists on living in Sendai. As do many elderly. They wish to return to their place, their home. Her daughter, on the other hand, insisted that she move to Tokyo to be with her family. Yukiko chose to live with her daughter and grandchildren, at the price of spending less time with her mother. She visits her mother in Sendai regularly. Yet the stress of being away from her and her constant travel can take its toll. Yukiko represents the anguishing quandary many displaced survivors face, particularly the elderly who desperately seek to return home. For many of them, home means going back to their place where they lived with family, friends, and neighbors. What is home? And what is the relationship between place and home? Yukiko, torn between being with her daughter and her mother, finds her home now with her daughter in Tokyo. Yet her home is also with her mother, who stays in her place in Sendai. Young adults find their place and their home where they can find work. That's their home. But it is not home for their parents and other family. For Yukiko, to be with her daughter is to be without her place with her mother. The elderly are thus doubly displaced. First forced to move away from their place and home. And now displaced from their children and families who are moving elsewhere, away from them.

We later discover our social geography, relationality, with our peers, our classmates, teachers, parents of our classmates, and so on. We discover what lies beyond our homes, new neighborhoods with its grocers, barbers, and families up the road, friends just blocks away. We constantly discover and stretch new social boundaries and new social territories beyond those boundaries. All this becomes our early community in which we find out that nothing happens in total isolation. Actions have their consequences and ripple effects. Simply put, we can't get away with much in our small world without being detected. And as we learn how much bigger the world really is, we also realize the same truth: all actions have their ripple effects, tangible or not. All happenings are somehow, strangely intertwined. We live in an inescapable sea of human interaction, mingling in, affected by, and bringing about ours and others' countless responses to this co-mingling, whether we desire it or not.

Here is an example of the power of our social climate. As we've seen above, Watsuji's notion of climate extends in all directions. It clearly encompasses the physical, those meteorological and geophysical forces we more naturally tend to associate with climate like weather conditions, where we live, and natural hazards we face depending upon where we live. Living by the sea naturally presents itself with a host of climatic features that we can embrace, be more indifferent to, and encounter as challenges. How we adapt, live in, and even survive our natural conditions depend upon how we brace ourselves for the hazards we will face. After all, it's our choice to live by the sea, for all sorts of reasons. Our choices have their consequences that will come at their own speed, consequences for which we may be prepared—the "northeaster" moving its way down from Canada coming our way, tide surges, floods, and those razor sharp winds—and those that blindside us. Such is the tempo of the sea, and we choose to be its neighbor.

So, our intuitive sense of climate echoes itself in those daily, hourly, now 24/7 weather reports that chime in our dwellings like white noise, the descriptions imbedding themselves into our everyday unconscious. However, something strange followed in the wake of that giant 3.11 earthquake. Even though it registered at magnitude 9, its colossal rupture surprisingly took less human lives than one would expect from such a powerful punch. Of the total fatalities, the quake itself inflicted less than 5 percent.[34] Strange, given its giant force. It does not equate with how we naturally link death and destruction with the degree of nature's force. We presume that the more powerful equals the more devastating equals the more destructive adding up to more injuries and deaths. The quake's punch, however, lasted long into the late rounds, and its power generated the real killer, a huge tsunami with waves that reached heights of ninety feet, easily overtaking miles of sturdy, and extremely costly and still controversial, seawalls all along the coastline. These waves just as effortlessly breached Fukushima Daiichi Nuclear Power Station's protective levees, knocking out backup generators and annulling reactors' cooling systems. We know the rest of the story—an unfolding tragedy of communicative *faux pas*, incompetency, and political and industrial ineptitude. The waves claimed thousands of victims, nearly 16,000, and still over 2500 still missing.

Given the tsunami's unexpected power, terrible loss of human lives in towns and villages along the coast makes perfect logical sense and reinforce our natural assumptions regarding climate. Namely, the degree of nature's power = the amount of human loss. But, even with the waves' undisputed dominance, something unexpected happened. Just about all towns and villages along the Sanriku seacoast experienced terrible destruction. However, there is an unevenness in the death rates, deaths per capita, along the coast.[35] Some towns and villages had a much higher death rate than other towns. Nothing unusual here. After all, those living in more protected inlets as opposed to facing open sea would naturally experience less deaths, owing to a geophysical arrangement that would reduce wave heights. Our assumption: the higher the waves, the more deaths. Moreover, living further inland, even though not on higher ground, would still be safer than living close to the sea. Again we assume: lower wave height would incur less deaths. We naturally associate wave height with its destructive force. And, in fairly populated areas, the higher the waves, the greater likelihood of human fatalities.

Normally, when we think of climate, especially in extreme conditions of disaster such as 3.11, we think of nature's magnitude, the unfathomable power of the sea and its waves, and the geography and physical setting toward which the waves steer. Faced with an oncoming tsunami, if we were to predict the degree of destructiveness and human fatalities, given our natural sense of what constitutes climate, we would associate destructiveness and death with wave height and force. But, this was not always the case in the 3.11 calamity. Some towns that faced very high waves actually suffered less human deaths per capita than towns engulfed by much shorter waves. In their eye-opening analysis, Daniel Aldrich and Yasuyuki Sawada describe how waves around sixty feet high engulfed the town of Miyako and waves less than 25 feet, still perilously high, drowned Higashimatsushima. Yet the smaller waves killed more people per capita in Higashimatsushima than the much larger waves in Miyako.[36] How could this be? Especially since seawalls were not significantly different in height. If, as Watsuji never tires of reminding us, we are imbedded in climate so that our lives are indelibly intertwined within climate and in-betweenness, there is another aspect of climate that we tend to overlook. Climate is social as well as physical. It is this aspect of climate, our social climate, which Watsuji underscores more philosophically in his *Rinrigaku*.

Let's apply social climate to disaster preparedness and response. Social climate involves human habitation including demographic features, size of population, numbers of vulnerable citizens, and so forth. Consider demographics. Many of these coastal towns have a high proportion of elderly residents. We therefore expect the fatality rates to be especially higher in towns where there are more elderly. The capacity to survive disaster rests very much upon the vitality and physical strength of towns' occupants. The 2004 Indian Ocean tsunami tragically confirmed this. With over 200,000 deaths, most fatalities were children, elderly, and women.[37] Children, elderly and other vulnerable groups are those who are most victimized in these natural disasters. Social climate also includes human interaction and struggle in facing nature's force. This social climate reveals itself in whether or not and to what degree we humans design our towns' physical infrastructure—well-signed escape routes, paved roads, preparedness plans, and education regarding the physics of earthquakes and tsunamis. Consider education. Another factor in that terrible Indian Ocean calamity was lack of education and awareness about warning signs. Many villagers, particularly tourists, failed to recognize that the receding ocean water immediately after the earthquake was a clear-cut sign of an impending ocean avalanche. Numerous victims walked far out in wonder at the receding sea.

Social climate thus relates to community networks, neighborhood dynamics, and community leaders and the degrees to which these networks, neighbors, and leaders engage in what Aldrich calls "bonding capital." "Bonding social capital" is a critical component of "social capital," or degree of social cohesion that certainly affects the sense of and commitment to community, sharing information, and working together in mutual assistance.[38] Earlier studies have shown this to be the case during the 1995 Kobe earthquake.

Then, the two neighborhoods of Mano and Mikura in Kobe were next to each other. However, though close in proximity, the two neighborhoods exhibited contrasting responses to the fires that began to rage throughout the city. The inhabitants in Mikura fled in disarray, without any organization, many of them in panic. Whereas Mano's neighbors managed to act together and organize their own fire brigades to battle the spreading flames.[39]

Another way of measuring "bonding capital" lies in ascertaining a community's crime rate. Aldrich and others, like Robert Putnam in his best-selling *Bowling Alone*, see a significant co-relation between a town's crime rate and its social cohesion.[40] That is, if crime rate is relatively high prior to any disaster, this signals a diminished sense of community that spurs the likelihood of crime in the aftermath of disaster. And particularly those who live alone suffer higher risk of death in the face of a disaster. Take the deadly 1995 Chicago heat wave, another insidious type of "wave," that killed over seven hundred people.[41] Many of those who died lived alone, most of them elderly who felt trapped in their apartments because they were afraid to venture out into the dangerous streets outside. Their neighborhoods had been riddled with crime for decades. Going outside would expose them to dangers of being victimized, mugged, even murdered. For many of them, they had no choice but to stay in their oppressively over-heated living quarters without air-conditioning. These victims lived isolated lives, disconnected from family and friends. Chicago's deadly heatwave illustrates their alienation in a dangerous city district, and how their estrangement ultimately leads to their sad, lonely deaths. Japan has a relatively low rate of crime, particularly in rural areas. According to Aldrich's studies, towns and villages along Sanriku with lower crime rates tend to better stand together as a community after a disaster. Furthermore, these same towns tend to have lower mortality rates.

> Holding all other factors constant, communities which had higher crime rates before the tsunami will experience greater levels of mortality during it. A municipality which had 15 crimes committed per 1000 people would be expected to have a death rate some 30 times higher than one where there were only one to two crimes per 1000 people.[42]

In any case, there is a strong relationship between crime and trust: high crime both affects and reflects degrees of public trust.

These ingredients of social climate help to explain the unevenness of the 3.11 tsunami's fatalities along the Sanriku coast. This is the in-betweenness that is life-saving. This is the *internal infrastructure* that is just as pivotal, if not more so, in surviving disaster than the *external infrastructure* of roads, fire department and hospital preparedness, and rehearsed evacuation tactics. And, as Aldrich and Sawada note, the degree of social cohesiveness rests upon degrees of mutual trust.

> Where communities share norms and have deep reservoirs of trust in each other and authorities, they are more likely to act on group information about risks. Therefore pre-disaster levels of *bonding capital*—the ties that bind residents to friends and neighbors—may matter because of their influence on information sharing, mutual help, and collective action.[43]

Trust is the oxygen of human interaction. And, as it becomes clear in accounts of survivors, it is the ground for resilience and hope, what I call "moral grit."

What is most important here is that *how people affected by the disaster already interact with each other as a community, with and for each other*, plays a crucial role and strongly influences death rates from the tsunami, more so than wave and seawall height. Without question, physical infrastructure plays a prominent role in mitigating the effects of a disaster event. Seawalls are critical in this respect. It was the failure of the seawalls in New Orleans that spawned such devastation more so than the hurricane itself. Throughout the Sanriku coastline, exorbitant sums of monies had been poured into constructing adequate seawalls for tsunami protection. At the same time, having relatively high, strong seawalls can also bring about a false sense of security for occupants who live behind these seawalls. Many of these people can feel safer than they are. Seawalls can thus be a type of "moral hazard," inducing a passive, naively trusting state-of-mind, one that can spell death.[44] Yet experience is always the better teacher. Those who survive seek shelter on higher ground, regardless of the levee.

Building community, strengthening social and bonding capital, cultivating connectedness—these are precisely the intended aims of Tono Magokoro Net, with whom I worked in the winter of 2014. Although we volunteers spent much of our time in physical labor shoveling snow and mud, chopping ice, moving rocks, digging hard ground for future gardens, splitting wood; duties also included working effortlessly to build community through special programs for children and elderly, those in shelters, and numerous efforts to integrate generations.

Let us now summarize the main points in this extended discussion of Watsuji and climate, for this constitutes the key theme throughout this book. We cannot escape our climate. Climate incorporates both physical forces of nature and social interaction. The tragedy of 3.11 illustrates the force of our physical climate—the power of the earthquake and ensuing tsunami, waves, weather conditions, geophysical factors such as exposure to the sea, length of coastline, and proximity to higher ground. It also demonstrates social climate—size of population, demographics and vulnerability of inhabitants, infrastructural features like access to roads, paved roads, evacuation routes, emergency planning, and firefighting and medical resources. Most importantly, social climate includes the key component of social cohesion and bonding capital.

Watsuji centers his ethics around our unavoidable imbeddedness within both types of climate. Moral behavior, how we live and act with each other is grounded upon our in-betweenness, our *aidagara*. Our in-betweenness is the space and condition for acknowledging our relatedness with the other. In which case, the other is not some "other," but someone with whom we share common ground, humanity, and destiny. Only through acknowledging this can we pour what St. Bonaventure calls, in describing the actions of the Good Samaritan (Luke 10:25-37), the "oil of compassion" onto the wounds of our neighbor.[45] Unless we acknowledge our imbeddedness in our social climate and cultivate social cohesion and genuine connectedness, when calamity strikes, whether individual or collective, small or large, we are less likely to find a hand that will reach out to us and help pull us through, like the mysterious hand that pulled Akiko Iwasaki out of the deadly churning waves. (We meet Akiko in chapter 4). Without social cohesion and bonding, the community is imperiled.

Here's the rub. *This social bonding must exist prior to the disaster.* When the waves sweep in unexpectedly, fear, dread, and panic can immediately overtake reason. Without a strong sense of community to help pull us out of our dread, without others to save us from ourselves, we more easily fall victim to a cold, dark, and quick fate in the waters that swallow us. In the long run, how we are as a community before the storm will determine whether and how we make it after the storm. This is a message we cannot afford to forget, for Nature will show her full force increasingly in the future, and with mounting intensity.

HOPE DIES LAST: MORAL GRIT

After World War II, there was hope for Japan despite near annihilation. Now, many claim that, as Murakami Ryū writes in his *Exodus of a Country of Hope*, hope is what people do not have.[46] For Murakami, people are so consumed by materialism, their interior life shrivels. What's left? 3.11's instant liquefying of coastal towns, nearby cities, homes, lives, families, and livelihood has scarred survivors with the rage of uncertainty, instability, and an ongoing threat of radiation poisoning. For them, the future is less probable. Uncertainty rules.

After 3.11, all is unhinged. Yet the pervasive mud in the aftermath of 3.11 covered a landscape already muddied. As Allison poignantly describes, this *ryūdōka*, or "liquidization or flexibilization of work and life," has been unraveling well before 3.11.[47] I see this *ryūdōka* in peoples' eyes when we meet while sharing our solidarity through stories. I see this sadness in my friend Hiroko Suzuki's eyes when we speak of the disaster. Hiroko lost her work in Sendai as her travel agency immediately shut down after the earthquake and tsunami blitzed the city. It flattened a good portion of her neighborhood and her mother's home. It also forced her to seek work further north in Tono. This in turn meant being far enough away from her mother to not watch after her.

The calamity of 3.11 pounded the nail deeper, nearly sealing the collective coffin completely shut. But not entirely. Light and air still manage to seep in. Can Japan free itself from being buried alive? That's what this book is about. There's light and a crack in the coffin.

For the first few weeks after 3.11, it was all a matter of bare survival, without running water, power, food, warm clothes, and blankets. In this struggle to eke out each day, there was also the internal battle against time. Here, each hour seemed endless, each day unending, each night an eternity. Ogyu Shumiyo is the head priest, or *jūshoku*, of a Buddhist temple at the edge of Otsuchi. When Hiroko, her friend, and I met Ogyu he shared with us his story and prayers of gratitude. His temple is situated on a relatively high road and just far enough inland from the shore so that when the tsunami hit the coastline, the train of waves reached as far as the steps leading to the temple without creeping inside. Neighbors frantically rushed to the temple for safety. In the following days, bodies had washed up to the temple steps. Miraculously, four of them were still alive. He and a handful of evacuees pulled the soaked bodies inside, dried them with towels—there was no heat—and managed to keep them alive. Altogether, the temple sheltered around forty surviving neighbors, all completely fatigued. The temple had no running water. The toilet was useless. Fresh water was crucial for washing and toiletry. Hygiene was a nasty issue with each day, and with forty people, water was essential. Each day, the *jūshoku* made it his mission to climb the steep hillside behind the temple and make his way to the community center located at the top of the hill overlooking Otsuchi's bay area. He brought with him two large empty buckets to fetch water from the community center. After filling the buckets, he gingerly carried them down the steep incline to the temple, a round trip of easily one mile. He made this journey at least three times a day. Forty survivors, including the priest, lived in that temple for six months, from March 11 to mid-September. The temple now houses hundreds of urns that hold the unidentified remains of victims.

Despite dark shadows, hope has not yet died, affirming all the more that *aida*, our in-betweenness, is inescapable. Post-war recovery demonstrated the triumph of stability even after being atomically jarred apart, instantly incinerated, and nearly obliterated. Ultimately, 3.11 reminds us that *aidagara* interrupts our plans and projects. Literally situated on a fault line, Japan, what appears to be one of the most stable countries in the world, will always be subject to its situatedness in-between human, institutional, economic and social order, and the vast rumbling underneath, with tremors that will routinely shatter the surface and countless lives.

Mexican-American farmworker Jessie De La Cruz (1919–2013), a farm laborer from the age of five, knew poverty and hardship first-hand. She joined César Estrada Chávez's (1927–1993) United Farm Workers Association (later United Farm Workers) and became its first female recruiter. She faced stubborn hurdles, yet never tired of working to help relieve the plight of exploited Mexican and Mexican-American farmers. She once said, "Whenever things are bleak and seem hopeless, we have a saying in Spanish, '*La esperanza muere al ultimo.*' Hope dies last."[48] Jessie remains an icon among Mexicans. Her resilience and life-long commitment shows how ordinary people can bring about extraordinary results.

"Hope dies last." There is hope precisely in our precarious in-between. The Buddhist priest and his temple survivors are living testament to this. This is a hope that erupts from a social defiance that Allison calls the "soul on strike." What is the source of this defiance? A compassion that grounds itself on acknowledging our indelible *aidagara*, our fundamental, intrinsic in-betweenness, realizing that all things are inherently interconnected. This is the teaching behind *pratityasamutpada*, a mutual rising and falling of all that is, in which each event is each other's cause and effect. This is a remembrance of the sea from which we come. All-powerful cultural forces impress upon us this inescapable relationality. Hope springs from recognizing *aida*, our inevitable in-between physical and social climate. This in turn inspires defiance and what I call "moral grit." Allison's "soul on strike" is this moral grit.

Connectivity Is Not Connectedness

Moral grit has to do with an attitude that spurs a living-out of genuine connectedness, one that is face-to-face, and eye-to-eye, not some online connectivity. This connectedness, or embodied presence to and for each other is now more vital than ever. Note Asahi Shimbun's citing of the June 11, 2011, survey in which over 70 percent of the respondents urged the need for such connectedness.[49] In this time of loneliness and *kodokushi*, "lonely death," with its prevailing uncertainty as to what the future harbors, according to the survey, these 70 percent feel it is crucial to cultivate one-to-one, face-to-face connections, *tsunagari*, that is, connectedness. This differs from connectivity, which has become our default way of connecting via digital and online devices. There is a crucial distinction between connectedness and connectivity. Connectedness is embodied and personal, face-to-face, engaged, and physical. Connectivity is connecting with others through devices and is screen-based. It is assuredly a powerful tool for connecting. However, this "connecting" remains disembodied. It lacks what we desperately need—physical connection and contact.

The young father Jin Imai barely escaped waves at his heels as he sprinted uphill for his life outside of Otsuchi. (We will meet him in chapter 5.) Jin realized the limits of connectivity soon after the disaster. After desperately searching for days for his wife and son and finally finding them alive, he learned that crisis, however difficult it is to endure, is also an opportunity for change, for genuine connectedness. He now aims to impress upon Japanese youth the idea that connectivity only goes so far. He repeatedly tells them that true community can only come about through personal connectedness, interpersonal engagement with and for each other.

My experience as a volunteer in Tohoku embodied for me this connectedness. None of us relied on our digital devices. We were all solely focused on doing whatever we could to help survivors in various coastal towns like Kamaishi, Otsuchi, and Rikuzentakata. Some of us did bring along our iPads and mobile phones, more to take photos of the stricken areas. My iPad was a way of Skyping in the evening or very early morning with my wife Brooke, thirteen hours away. But even with the taking of photos, we were extremely sensitive to the fact that the ravaged coastline where we worked was also sacred ground, in effect a mass killing field and "burial" site where thousands of helpless victims met their fate, pulverized by sea and debris, and incinerated by raging fires. We worked where people died violent deaths. We honored these coastal sites, prayed reverently at memorials, and only took photos once we were given explicit permission from our group leader for that day. And even when we did take photos, we knew in our heart of hearts that we could never capture the reality. Figure 2.1 shows the memorial site overlooking a still devastated Otsuchi, three years later.

The essential ingredient in moral grit involves attitude. Attitude. When the well runs dry, when things are at their lowest ebb, all seems hopeless and helpless. Hope can easily turn into despair. Moral grit transcends the growing dark-age mind-set whereby healthy skepticism sours into cynicism, when everything becomes subject to criticism, with little in the way of helpful advice. When temperaments become weary and worn-out, and just about all of life takes on a dourness, and one's days are wrapped in distrust. When we paint our daily canvas with the colors of sarcasm, scorn, and suspicion. Moral grit wants nothing to do with that. It defiantly refuses to fall into the pit of despair. Moral grit demands the ability, at times extraordinary, to sustain hope through the darkest of times. Without it, we slip into utter darkness, an abyss. My parents are intimate examples of moral grit. Dad, an Irish-American GI, stationed in Kyushu, met Mom during the Korean War. Defiance is pretty much what they embodied together in the face of one setback after another, from Mom's forced exile from her Japanese family for

Figure 2.1. Memorial overlooking Otsuchi, 2014

marrying a *gaijin*, to Dad's "punishment" by the military for marrying a "Jap," to their later exile to our all-Irish Catholic neighborhood in a small, provincial New England town.

Throughout its history, Tohoku and its Sanriku coast has endured a rough legacy of famines, earthquakes, tsunamis, and droughts. Their past has thickened peoples' skins and has contributed to their bouncing back after each plight with an even more resilient spirit. It is their moral grit that enables this. Moral grit is their defiance in the face of adversity. A popular expression among people in Tohoku is *Maketara-akan, tomo ni yaru shika nai*—"We refuse to lose, we must get through it together."

NOTES

1. Norimitsu Onishi, "Death Reveals Harsh Side of a 'Model' in Japan," *New York Times,* accessed Dec. 19, 2015, http://www.nytimes.com/2007/10/12/world/asia/12japan.html?hp=&_r=0.
2. Anne Allison, *Precarious Japan* (Durham, NC: Duke University Press, 2013), 19.
3. "Suicide in Japan," *Nippon.com* (Oct. 9, 2014), accessed Jan. 15, 2015, http://www.nippon.com/en/features/h00075/.
4. Allison offer a superb analysis in her *Precarious Japan.*

5. See John Dower's brilliant analysis, *Embracing Defeat: Japan in the Wake of World War II* (New York: W. W. Norton & Company, 1999).

6. Allison, *Precarious Japan*, 5.

7. Allison, *Precarious Japan*, 5. Clearly, the shift is global. Norway, a welfare country with its 4.5 percent poverty rate now faces challenging immigration issues. For instance, how long can it sustain a mounting number of Somalis who do not work and yet are cared for by the system? Personal communication, Professor Julie Feilberg, University of Trondheim, 5.13.14.

8. Allison, *Precarious Japan*, 16.

9. Michiyo Nakamoto, "Plight of Fukushima farmers takes it's toll," *ft.com*, Japan, accessed Dec. 19, 2014, http://www.ft.com/cms/s/0/5df7847e-5a07-11e0-ba8d-00144feab49a.html#axzz3PwApkjh4.

10. Robert Gilhooly, "Suicides upping casualties from Tohoku disaster," *Japan Times*, June 23, 2011, accessed Dec. 19, 2014, http://www.japantimes.co.jp/news/2011/06/23/national/suicides-upping-casualties-from-tohoku-catastrophe/#.VJSnq8BEA.

11. Lucy Birmingham and David McNeil, *Strong in the Rain: Surviving Japan's Earthquake, Tsunami, and Fukushima Nuclear Disaster* (New York: Palgrave Macmillan, 2012), 157.

12. Interview with Anne Cecilie Fossum, Oslo, June 21, 2012.

13. Senter for Krisepsykologi, "Etterlatte foreldre, partnere, søsken og venner etter Utøyadrapene 22.07.2011: Oppsummering av forskningsprosjekt–2¼ är etter," Bergen (2014). [Center for Crisis Psychology, "Bereaved parents, partners, siblings and friends after Utøya murders 22.07.2011: Summary of a research project—2¼ years later," Bergen (2014)].

14. Kenzaburo Oë, *A Personal Matter*, trans. John Nathan (New York: Grove Press, Inc., 1969).

15. See Soda Kazuhiro's commentary, "*Seishin* (Mental)," *Berlinale* (2009): 140-43, accessed February 8, 2015, https://www.berlinale.de/external/de/filmarchiv/doku_pdf/20091068.pdf.

16. Birmingham and McNeil, *Strong in the Rain*, 161.

17. See Zygmunt Bauman, *Culture in a Liquid Modern World,* trans. (from Polish) by Lydia Bauman (Cambridge, UK: Polity Press, 2011).

18. See Anthony Giddens, *Runaway World: How Globalization is Reshaping Our Lives* (New York: Routledge, 2003).

19. Jonathan Lear, *Radical Hope: Ethics in the Face of Cultural Devastation* (Cambridge, MA: Harvard University Press, 2006); also cited in Allison, *Precarious Japan*, 193.

20. Steve Odin, *The Social Self in Zen and American Pragmatism* (Albany, NY: State University of New York Press, 1996), 52.

21. I introduce Watsuji Tetsuro in my *The Pulse of Wisdom: The Philosophies of India, China, and Japan*, 2nd ed. (Wadsworth Publishing Company, CA, 2000; 1st ed., Belmont, CA: Wadsworth Publishing Company, Inc., 1995), 276–77, 318–20, 149–50.

22. Watsuji Tetsuro, *Fudo: Climate and Culture: A Philosophical Study*, trans. Geoffrey Bownas (Westport CT: Greenwodd Press, 1988; reprint, orig. publisher Tokyo: Monbusho, 1961), 6, 16. It was originally published in 1935, taken from notes Watsuji had written in the 1920s. It was translated into English in 1961.

23. A. Kohut et al, "Economy, Jobs Trump All Other Policy Priorities" (Washington, DC: Pew Research Center for the People and the Press, 2009). The priorities were ranked in this order: economy, jobs, terrorism, Social Security, education, energy, Medicare, healthcare, deficit reduction, health insurance, poverty, crime, moral decline, military, tax reductions, environment, immigration, trade, and global warming. Cited in Heidi Cullen, *The Weather of the Future: Heat Waves, Extreme Storms, and Other Scenes from a Climate-Changed Planet* (New York: Harper Collins Publishers, 2010), 5.

24. See discussion in Cullen, *The Weather of the Future*, 6ff.

25. See Michael C. Brannigan, "Guarding against a deadly outbreak of unfounded fear," *Albany Times Union*, Nov. 26, 2014, http://www.timesunion.com/opinion/article/Michael-Brannigan-Guarding-against-a-deadly-5920030.php.

26. Nassim Nicholas Talib, *The Black Swan: The Impact of the Highly Improbable*, 2nd ed. (New York: Random House, 2010).

27. Watsuji Tetsuro, *Rinrigaku*, trans. by Yamamoto Seisaku and Robert Carter (Albany, NY: State University of New York Press, 1996; orig. published in Japanese by Iwanami Shoten, Tokyo, 1937, 1942, 1949).

28. Robert E. Carter, *Encounter with Enlightenment: A Study of Japanese Ethics* (Albany, NY: State University of New York Press, 2001), 126.

29. Carter, *Encounter with Enlightenment*, 127.

30. Michael C. Brannigan, *Cultural Fault Lines in Healthcare: Reflections on Cultural Competency* (Lanham, MD: Lexington Books, Rowman & Littlefield Publishing Group, Inc., 2012).

31. Emmanuel Levinas, *Humanism of the Other*, trans. Nidra Poller (Urbana: University of Illinois Press, 2006), 31–32.

32. Carter, *Encounter with Enlightenment*, 131.

33. Nikkō, which means "sunshine," is a beautiful mountain city in Tochigi Prefecture and one of Japan's most popular tourist destinations with its magnificent shrines and temples. Historically, it is a sacred site since the eighth century when a Buddhist priest established a temple of the famous Buddhist Tendai school. It is where the imposing seventeenth century Tōshōgū Shrine contains the mausoleum of the Shogun Tokugawa Ieyasu who managed to conquer and unite all of Japan and also where the less lavish Taiyūin Reibyō houses the remains of his grandson Tokugawa Iemitsu. For Tsurukame Daiichi, see http://www.nikko-turukame.jp/.

34. "Damage Situation and Police Countermeasures," National Police Agency of Japan, accessed Dec. 15, 2014, http://www.npa.go.jp/archive/keibi/biki/higaijokyo_e.pdf.

35. Daniel Aldrich and Yasuyuki Sawada, "The physical and social determinants of mortality in the 3.11 tsunami," *Social Science and Medicine*, 124 (January 2015): 66–75. They offer an excellent, quantitative study of the surprisingly variable mortality rates from the tsunami.

36. Aldrich and Sawada, "The physical and social determinants," 67.

37. Elizabeth Frankenberg, Thomas Gillespie, Samuel Preston, Bondan Sikoki, and Duncan Thomas, "Mortality, the Family and the Indian Ocean Tsunami," accessed January 15, 2015, http://eml.berkeley.edu/~webfac/alexroth/dthomas.pdf. The authors here give over 160,000 deaths. Most sources cite higher figures.

38. Daniel P. Aldrich, *Building Resilience: Social Capital in Post-Disaster Recover* (Chicago, IL: University of Chicago Press, 2012), 31ff.

39. Aldrich, *Building Resilience*, 73.

40. Robert Putnam, *Bowling Alone: The Collapse and Revival of American Community* (New York: Simon and Schuster, 2000).

41. Eric Klinenberg, *Heat Wave: A Social Autopsy of Disaster in Chicago* (Chicago, IL: University of Chicago Press, 2002). See interview with Klinenberg, accessed Jan. 14, 2015, at http://press.uchicago.edu/Misc/Chicago/443213in.html.

42. Aldrich and Sawada, "The physical and social determinants," 72. Aldrich and Sawada discovered another interesting measure of social cohesion in the form of bonding capital – the relationship a community has to the ruling political party, in this case the Liberal Democratic Party (LDP). This relationship, that is, whether or not the majority of the population voted in favor of the LDP, is such that it will determine the level of support for and investment in local initiatives concerning infrastructural disaster prevention and mitigation plans.

43. Aldrich and Sawada, "The physical and social determinants," 68.

44. Aldrich and Sawada, "The physical and social determinants," 67].

45. St. Bonaventure, *De sex alis seraphim* [On the Six Wings of the Seraphim], trans. Sabinus Molitor (St. Louis: B. Herder, 1920), 30, cited in Daniel P. Sulmasy, *The Rebirth of the Clinic: An Introduction to Spirituality in Health Care* (Washington, DC: Georgetown University Press, 2006), 65.

46. Murakami Ryū, *Kibō no kuni no ekosodasu* [The Exodus of a Country with Hope] (Tokyo: Bunshun Bunko, 2002); cited in Allison, *Precarious Japan*, 6.

47. Allison, *Precarious Japan*, 7–8.

48. Studs Terkel, "Hope Dies Last," in *The Life of Meaning: Reflections on Faith, Doubt, and Repairing the World*, ed. Bob Abernethy and William Bole (New York: Seven Stories Press, 2007), 105.

49. Allison, *Precarious Japan*, 20.

Chapter Three

Who Is Our Neighbor? The Tono Lifeline

Honshu, the largest and most populous of Japan's four islands and considered Japan's mainland, lies between Hokkaido in the north, Shikoku in the south, and Kyushu in the southwest. Tohoku is the northeast region of Honshu. Tohoku has six prefectures: Aomori, Akita, Iwate, Yamagata, Miyagi, and Fukushima. Iwate, Miyagi, and Fukushima are the three heavily hit areas from 3.11. Wedged in between north and south, Tohoku's climate is unforgiving, regularly issuing bitter cold and unending winters, relentless winds, and oppressive summer heat. Tohoku's harsh mountain climate reinforces a need for solidarity against the elements. I can never forget its nonstop cold and wind as our small band of volunteers huddled in our quarters during the coldest part of the year in mountainous Tono, in Iwate Prefecture.

Tohoku's geography and topography clearly impact its lifestyles, worldviews, values, beliefs, and legends. In addition to its mountain climate, its proximity to the sea plays a powerful role. The Pacific Ocean washes up against Tohoku's eastern side, its coastal area called Sanriku. The sea harbors life-giving resources, with an abundance of fish and seaweed. The sea can also rage unexpectedly and consume entire villages, just as its fierce currents can pull swimmers out to drown. As with the tides, the sea gives and takes, though its taking is sudden and often unanticipated. The sea is a living force that marks the boundary between land and the beyond, between solid, dry earth and the vast unknown. For many Japanese, especially in Tohoku, the sea denotes the border between two worlds—this one and another, ever-present mystery.

For the people of Tohoku, mystery surrounds the sea. Ever since 3.11, there continue to be reported accounts of unexplainable occurrences by the sea outside of Minamisoma. There, roads were repaired and traversable a few months after the impact. Despite these repairs, motorists reported driving near the ocean and suddenly hitting an unseen object. When they stepped outside of their cars to investigate, they found nothing—no evidence or hint of any collision. Yet they were absolutely convinced that something was there that they ran into. Local police claim this was reported a number of times.[1]

We see this mysterious role of the sea and rivers in many Tohoku legends. The most well-known legends in Tohoku are about *kappa*, mischievous water-creatures who inhabit rivers. Known to be elusive, each *kappa* is the height of a small child with a turtle shell on its back. On the top of its head is a dish-like depression for carrying water. *Kappa* need water to survive. Usually green, in Tono they are red-faced. In his famous collection of folktales, *Tōno Monogatari*, Kunio Yanagita writes:

> In other areas, the *kappa*'s face is said to be green, but in Tono the *kappa*'s face is red. When she was young, Kizen Sasaki's great-grandmother was playing with friends in the garden when she saw a boy with a dark red face behind three walnut trees. It was a *kappa*. Those big walnut trees are still there. The area around the house is now filled with walnut trees.[2]

According to legend, many *kappa* live at Kappabuchi (the *kappa* pool), on the outskirts of Tono. I visited the pool during one of my travels there, first making my way through Denshoen Park and viewing its impressive collection of one thousand Oshira-sama, or gods of agriculture, then walking through neighboring farmland, and finally arriving at a swift stream and pond. But I found no *kappa*, though they may have been secretly watching. *Kappa* are known to have pulled unsuspecting children into the river to drown. They also occasionally impregnate women and kidnap children. Yanagita writes:

> Many *kappa* (mischievous water spirits) live in rivers. There are large numbers of *kappa* in the Saru-ga-ishi River. In a house by the river in Matsuzaki village, women have become pregnant with a *kappa*'s children for two generations. When the *kappa*-children are born, they are hacked into pieces, put in small wine casks, and buried in the ground. They are grotesque.[3]

Kappa can also protect and give special blessings. Frightening in appearance, they scare children away from dangerous places, like rivers and rocks. And if a pregnant mother prays at Kappabuchi, she will have enough healthy breast milk. When I met Tono's mayor, Toshiaki Honda, he awarded me with a "kappa catching license." When I asked him if anyone has captured a *kappa*, he simply shook his head "no."

TOHOKU'S CLIMATE AND HISTORY OF AFFLICTION

The Tohoku region, perhaps Japan's most traditional area where countless visitors throng from all parts of Japan and overseas because of its sustained and rich reliance on time-honored customs, festivals (*matsuri*), and rituals, also has an enduring history of poverty and deprivation. This is evident during terrible famines which took their toll on rice farmers. Tohoku is known for its rice farming and, prior to 3.11, rice from Tohoku amounted to 25 percent of Japan's total rice output.[4] Unpredictable weather along with heavy taxes levied on the farmers occasionally brought about severe downturns on rice farming. Extreme, fickle weather has set in motion major famines in Japan. One of the most devastating was the 1732 Kyoho famine in northern Kyushu which took the lives of nearly 20 percent of the population in the Fukuoka region, around 15,000 people. (This is where my mother, her family, and I were born, and where my uncle Fujiya, Aunt Shizuka, and cousins still live.) At that time, Japan's entire population was 25 to 30 million.[5] Fortunately, there were no major famines in that region since.

Tohoku, however, has had a sustained history of terrible famines. In 1783, following intense floods, volcanic eruptions from Shima Island and Mount Sakarujima near Kagoshima and Mount Asama, excessively cold summers, and unremitting rain, the Tenmei famine lingered on from 1782 to 1785. According to a samurai observer:

> The loss of life through starvation in the northern provinces [Tohoku covers most of the northeast of Honshu] was dreadful. There was nothing to eat but horse-flesh or, when this ran short, dogs and cats. Once these were consumed, people died of sheer starvation in great numbers. In some villages of thirty, forty, or fifty households not one person survived, and nobody can say who had died or when, for the corpses were unburied and had been eaten by beasts and birds.[6]

There remains dispute over the number of deaths, ranging from 6,000 to 64,000. One figure goes as high as 260,000 deaths.[7]

From 1836 to 1838, an aberrantly cold spring, cold and wet summer, and a far-too-early snow severely spoiled rice, barley, and wheat crops, yielding instead the infamous Tempo famine. Again, the Tohoku region was the worst hit. There was mass suffering and starvation. Deaths are estimated to run from 20,000 to over 100,000, but for the same reasons above, are impossible to confirm.

"A Community of Skeletons"

Then we have the Great Famine of Northern Japan in 1905–1906, or Tohoku Famine (*Tōhoku Kikin*), sometimes called "the Great Tohoku Crop Failure," *dai Tōhoku kyōsaku*. Tohoku's Sanriku coast was still reeling from the June 15, 1896 Meiji-Sanriku earthquake and tsunami that took over 20,000 lives and left over 60,000 homeless. In 1905, it was uncharacteristically wet and cold with little sunshine throughout the summer. Temperatures fell around 4.8 degrees Celsius below average.[8] Fall witnessed an extreme change with excessively high temperatures. This spelled doom for rice crops and brought about one of the worst crop failures in recent times for Tohoku. Among all prefectures, there was meager rice to harvest, and this devastated the region. The famine was so bad it made international headlines like the *Auckland Star*'s "The Famine in Japan: Selling the Children for Money," the *Los Angeles Herald*'s "Japan Famine Most Terrible of Late Years: One Million Estimated to be Starving," and the *Pittsburgh Gazette*'s "Rice Famine in Japan, 650,000 Now Eat Straw."[9] Though some headlines like the selling of children were grossly exaggerated, the suffering was real and widespread. People throughout Tohoku, reduced to malnourishment and ruin, were scrapping for food in the forests, nuts, berries, even tree bark. Nature is cruel. An extremely harsh winter with heavy snows, one of Tohoku's worst winters on record, added to the untold misery. In February 1906, the *Christian Herald*, in a plea for international aid, writes:

> People who live in lands of plenty, where such a condition of things is unknown, can form no idea of the character of the suffering, nor of the straits to which these poor villagers have been driven, to keep the life within their miserable bodies. They are for the most part *like a community of skeletons* . . . It is impossible to describe the situation in these village homes of the simple Japanese farmers.[10] (author emphasis)

One-third of Tohoku, one million out of a total population of three million, was poverty stricken. What is interesting is that although plenty of aid did arrive from various international sources, Japanese authorities preferred to sustain an ethos of self-help and self-reliance among the Tohoku people rather than depend on charity. After the American Red Cross sent its funds, the Japanese Red Cross responded: "Before distributing the money which you sent to us, we very carefully investigated the best methods by which no peasant there should 'laze' away his time, simply relying upon such help."[11] Here, we see a stress on maintaining strong stoicism, resilience, and interdependency among Tohoku people. This same spirit of stoicism played itself out in the aftermath of 3.11.

During this famine, it also became increasingly apparent that many who lived outside of Tohoku perceived the people of Tohoku as being less intelligent and rather backward. The people in Tohoku were certainly poorer. But the famine seems to have spurred a growing stigma labeling them an "inferior class." Here is a statement from a foreign missionary conducting relief work in the region.

> The famine is limited to about 500,000 farmers who seem to be unable to do any but farming work and those farmers, physically and intellectually, are an inferior class, even amongst their own surroundings. About 10,000 of the more active and intelligent have emigrated to the Hokkaido [Japan's northernmost island], and those who have remained are inferior people and certainly very much inferior to the people in the south. [12]

This attitude, sadly, continues to reinforce unfair stigma, stereotyping, and discrimination towards Tohoku, branded by some as Tokyo's "colony." Noted Tohoku folklorist Norio Akasaka uses this term to criticize Tokyo's historic exploitation of the people of Tohoku. [13] Akasaka directs the Yanagita Kunio Museum in Tono, a wonderful resource for studying the history of Tono, its surrounding region, and Tohoku, or Tohoku-gaku, "Tohoku studies." He situates the 3.11 calamity and its aftermath within the context of Tohoku's long-standing history of exploitation by Tokyo.

Japan's Colony

This idea of Tohoku as Japan's "internal colony" has deep roots. In the effort to construct a railroad through Tohoku between 1881 and 1890, though it was seen as a step towards modernization for Tohoku, critics believed the project further bled the region to serve the interests of Tokyo and other major cities. [14] In his 1920 history of Japan, Katsurō Hara further relates the marginalization of Tohoku to Japan's earlier seemingly inbred political striving to build its empire:

> The principal cause of the retardation of progress in [Tohoku] lies . . . in the fact that it is a comparatively recently exploited part of the Empire. Since the beginning of historic times, the Japanese have pushed their settlements more and more toward the north, so that the population in those regions has grown denser and denser. If this process had continued with the same vigour until today, the northern provinces might have become far more populous, civilized, and prosperous, than we see them now. Unfortunately for the North[east], however, just at the most critical time in its development, the attention of the nation was compelled to turn from inner colonization to foreign relations. [15]

Norio Akasaka points out that this mind-set of Tohoku's dependency upon Tokyo still prevails, not so much among people in Tohoku but elsewhere throughout Japan. For instance, as a way to coordinate response to the 3.11 disaster, the government established a Reconstruction Design Council on which he served. The Council hailed Tohoku as Japan's "manufacturing hub," in rhetoric recognizing Tohoku's progressive status. Yet, in reality, Tohoku workers are still exploited. In an interview with the popular newspaper *Asahi Shimbun*, Akasaka comments, "For instance, in a tsunami-stricken municipality, the operator of an auto-parts factory is forced to work for an hourly payment of 300 yen ($2.94), well below the minimum wage."[16]

Tohoku continues to supply manufacturing, rice, and electricity to Tokyo. A sad example of the latter is the Fukushima Daiichi Nuclear Power Plant that provides its electrical power to Tokyo, but does not serve Tohoku, other than offering employment. Akasaka relates this exploitation to Tokyo's upcoming Summer Olympics.

> Nuclear power plants are more explicit symbols of the "colonial" rule . . . During the campaign to bring the 2020 Olympic Games to Tokyo, the president of the bid committee emphasized Fukushima's recovery but bluntly added that Tokyo is safe because it is 250 kilometers [155 miles] away from Fukushima. To his eyes, Tohoku is a remote and peripheral region.[17]

In summary, all this shows how Tohoku embodies the notion of *aidagara* through its geography, climate, and history. It vividly shows that unavoidable relationship Watsuji describes between climate and its force on history.

> It is in our relationship with the tyranny of nature that we first come to engage ourselves in joint measures to secure early protection from such tyranny. The apprehension of the self in climate is revealed as the discovery of such measures; it is not the recognition of the subject . . . The activity of man's self-apprehension (man, that is, in his dual character of individuals and social being) is at the same time of a historical nature. Therefore, climate does not exist apart from history, nor history apart from climate. This can only be understood from the fundamental structure of human existence.[18]

In so doing, this climate and history of struggle has helped shape the resilience of the people of Tohoku. It vividly shows how climate imprints itself upon people's worldviews, beliefs, and values. Its climate and history has carved out in inhabitants a strong sense of being-together-through-it-all, a spirit of uncommon solidarity. Its harsh climate and history molded inhabitants who cannot help but brave it out together in community.

To be sure, Tohoku is not one homogenous culture. It represents a variety of cultures from different prefectures, each with its own unique history, cultural emphases, and traditions. Generally economically depressed relative to other regions, Tohoku is also less known, though some cities like Sendai stand out. Sendai, the capital of Miyagi Prefecture, is the largest city in Tohoku with a population over one million. It is known throughout much of Japan because of its oysters, shark fin, and *magura* (tuna) served in numerous restaurants. Sendai's popularity is due to how it serves the interests of major cities, particularly Tokyo. In contrast, at least prior to 3.11 and the nuclear disaster, Fukushima City was little known, even though the city, capital of Fukushima Prefecture with a population under 300,000, prides itself on its rich fishing industry. Compared to Sendai, historically, many parts of Fukushima were economically depressed and poor.[19] In earlier times, because of its poverty and cultural "backwardness," Tohoku was viewed by Tokyo and much of Japan as *Ezo*, meaning the "outer world." Now, Hokkaido is Japan's *Ezo*.[20] In the opinion of many, Tohoku remains that of Tokyo's "colony," viewed by many in Tokyo and elsewhere as still stuck in its past traditions and not in pace with the rest of modern Japan.[21]

THE GIFT OF TONO

Tono is located inland in the southeast of Iwate Prefecture, just over four hours by train from Tokyo. It is in the northern part of Honshu, west of the Sanriku coast. Known for its horse breeding, hunting, and farming, as of February 1, 2014, its population is 29,367, with 10,845 households. It covers 852 square kilometers, or 529 square miles, more area than Tokyo. The key to understanding the significance of Tono lies in its location. Its location between sea and mountains has proven vital regarding trade and disasters. It connects the coastal area with the inland. Coastal cities and towns of Otsuchi, Yamada, Miyako, Kamaishi, Ofunato, and Rikuzentakata are within a 50 kilometers (31 mile) radius of Tono. Tono itself has weathered numerous earthquakes. But its in-betweenness as bridge between the coast and mountains naturally protects it from tsunamis. And as this bridge, it has aided coastal areas during crises.

The June 15, 1896, Meiji-Sanriku earthquake and tsunami, whose epicenter was close to the most recent one in 2011, took over 20,000 lives and left over 60,000 homeless. One day after the earthquake, Tono immediately dispatched its people to the coastal areas to provide emergency relief with monies, manpower, work horses, food, clothes, and water. Tono citizens were actively involved in rebuilding coastal towns. On March 3, 1933, a huge 8.4 magnitude earthquake at nearly the same epicenter as the earlier Meiji-Sanriku quake and tsunami spawned a tsunami with waves as high as 28 meters, around 94 feet. This Sanriku Earthquake ravaged a good portion of the coastline, especially the towns of Ofunato and Taro, part of Miyako City, and took over 3,000 lives. Again, Tono offered invaluable aid to survivors.

Tono-shi, or Tono City proper, is situated in a basin by the Sarugaishi River, a valley surrounded by mountains, the Kitakami range. Three mountains in particular stand out: Mt. Hayachine-san (1,914 meters, or 6,200 feet), Mt. Rokkoshi-san (1,294 meters, or 4,200 feet), and Mt. Ishigami-san (1,038 meters, 3,400 feet). "*San*" underscores their position of honor. Tono-shi is where we volunteers stayed at its headquarters on the edge of town. (More about that in chapter 6.) The city's mountain climate exposes inhabitants to extreme weather. In winter, incessant snow, wind, and rain is that force of Nature which compels inhabitants to live and struggle together. As Mayor Toshiaki Honda puts it, "it is part of our DNA to help each other."[22]

Tono takes its folklore seriously. This "City of Folklore" is known throughout Japan for its local legends and traditions, like the *kappa* described earlier. These legends are immortalized in Kunio Yanagita's celebrated folktales, *Tōno Monogatari*, written in 1910. Yanagita modeled his tales on *Grimm's Fairy Tales* which was published in 1812 as *Kinder-und Haus-Märchen*. From 2010 to 2012, the city hosted special events to celebrate the 100th anniversary of *Tōno Monogatari* and the 200th anniversary of *Grimm's Fairy Tales*. Kunio Yanagita's collection comes mainly from tales told to him by Tono's young writer and folklorist Kizen Sasaki (1886–1933).[23] In Tono, *kappa* are everywhere. As soon as you step out of the train station in the center of town you can't help but see statues of *kappa* in a pond. The police box near the station is in the design of a *kappa*. There are *kappa* toys, souvenirs, and t-shirts. Road workers everywhere use *kappa* as work area markers. All throughout the stricken coastal areas, construction and repair sites are cordoned off by rows of these *kappa*. Indeed, Tono's official mascot is a *kappa* named "*Karin-chan*." *Ka* is short for *kappa*, and *rin* stands for the bellflower which *Karin-chan* carries, or *rindo*. And again, Kappabuchi, a popular site for tourists, is where the popular mythical *kappa* hang out.

I have made two separate visits to Tono after 3.11. The first was in the winter of 2012, and more recently in the winter of 2014. I was able to make these visits through the generous guidance and support of Kunio Aoki, East Japan Railways's Culture Foundation manager. Ever since meeting Kunio and his wonderful family when I lived in Westchester, New York, we've kept our friendship alive. When I shared with Kunio my thoughts of meeting with survivors, he offered his usual wise counsel. He and his expert assistant Mizuki Naito graciously accompanied me throughout my visit to the stricken

Tohoku region and Sanriku coast. Mizuki possesses that keen ability to discern the heart of our conversations. In view of the Tohoku dialect, she offered invaluable translation and interpretation in measured fashion.

For both visits, Hiroko Suzuki was my key intermediary and contact in Tono. After the tsunami destroyed her travel agency in Sendai, Hiroko was one of numerous women who lost their jobs after the impact. Tohoku had retained much of the strict traditional gender roles which meant that working women were usually dependent upon local employment and often worked part-time. Because 3.11 destroyed numerous local businesses, women were the first to lose work. They also found it very hard afterwards to find work particularly since the government generated much post-3.11 work in construction with jobs more suitable for men. This may help explain an increase in female suicides in the first seven months following the disaster, whereas according to some reports male suicide rates actually decreased during the 24 months post-disaster.[24] Hiroko, however, is of strong ilk and character. She possesses a quiet strength. And, through her many friends and connections in Tono, she was appointed director of the well-known Tono Furosato Village. Furosato is a reconstructed traditional village situated just outside of the city in a beautiful setting, surrounded by stunning mountains.[25] It preserves the area's precious legacy of *Nanbu Magariya*, L-shaped homes where people and horses lived under the same roof during the Edo period. Horses have always been a special part of Tohoku history. Hiroko loved her work there. She and all her staff were called *maburitto*, or "protectors" of the village and its lifestyle. I will always remember our visits to the village and Hiroko's introducing me to that old and cherished way of life and to her favorite horse, *hiroi yuki*, or "snow white." Hiroko now works as a manager in Tono at the Tono Travel Agency (Tonojikan), located across the street from the train station.[26]

Hiroko is gracious and kind in every sense. We've had so many conversations, and I feel as if she is my sister. She has a special bond with Nature and possesses the uncanny ability to see through apparent complexity and get to the heart of things. She was intensely interested in learning more about Henry David Thoreau (1817–1862). We made a pact that she would read more of Thoreau, and I would read more of Miyazawa Kenji (1896–1933), Tohoku's literary icon. She also loves Rachel Carson (1907–1964). Hiroko is tuned into traditional Japan. The popular Rokkoshi Shrine lies just outside of Tono, near the top of Rokkoshi Mountain. Hiroko once drove to the mountain site during my 2014 winter stay. Deep snow covered the road to the mountain, barely passable, but she made the trek up the mountain by foot to the shrine to offer prayers to the mountain deities for Brooke and me. Brooke and I keep her *omamori* gift, a token for good luck that she gave us from her visit to the mountain, in a special spot in our home.

Back to Tono and its climate. Tono's climate has indelibly influenced its sense of community among the people. Almost all inhabitants have strong family relationships, and many homes have different generations living together. Traditional rituals act as glue for this solidarity. Ceremonies like weddings and funerals are done formally and the citizens support each other. People also cherish their many regional dances. Older generations in particular value these dances and strongly insist on passing these dances on to the children and teaching them. Here, grandparents teach these dances to two and three year-olds.[27] In all, Tono and other towns throughout Tohoku have a rich appreciation of ritual, tradition, and the performing arts. Tono's Citizens' Opera has a forty year history. Locals stage and perform the operas, and the mayor himself participates in these. High school and junior high school students play in the orchestra. During Tohoku's many hard times, these energize and help strengthen people to survive hardships and challenges. In many ways, Tono embodies the spirit of Tohoku in priding itself on preserving traditional rituals and festivals, *matsuri*. These remain a vital glue to sustain the communities during poverty, famine, and other afflictions.

Impact and Response

On March 11, 2011, at 2:46 p.m., as the largest recorded earthquake in Japan's history struck the Tohoku region, citizens felt the ground shake like never before. Tono experienced the earthquake at magnitude 5 on the Japanese scale. Roads cracked wide open, buildings collapsed, including the city office, shattering the streets with debris. Citizens immediately ran to dedicated evacuation shelters. For days afterwards, there was no electricity and a huge shortage of fuel.

Timing is everything. In coastal towns, the timing was bad. There, schools were ready to let out their students, and mothers and bus drivers were already on their way to pick up children. In Tono, this was more of a case of better timing. The earthquake hit hard. But in the mountains, and forty kilometers from the coast, citizens were not likely to face a tsunami. The quake occurred in the afternoon when employees were still working in the city office. Once the earthquake subsided, city officials could therefore directly take action. At three in the afternoon, just 14 minutes after the earthquake, they ordered that the Tono Sports Park be opened to receive and coordinate rescue, recovery, and relief efforts, as in prior drills. Thanks to the leadership of its mayor, Toshiaki Honda, well before the disaster, Tono formed a coalition with other cities in the coastal areas in planning a strategy for dealing with earthquakes and ensuing tsunamis. Tono, along with Japan's Self Defense Forces (SDF), regularly practiced rescue and recovery drills to deliver aid and relief particularly to coastal regions. During this time, most Tono citizens were truly in the dark. Faced with their own pressing concerns—loss of heat, fractured roads, the threat of falling objects and landslides, and, of course, fires—the quake knocked out power and all lines of communication. They had no idea about the devastation on the coast. Though they knew that a tsunami was likely to follow the massive quake, they were completely in the dark as to this tsunami's colossal power and destructive force.

Remember transistor radios? Believe it or not, they were the only source of information after the earthquake. A few city officials including the mayor owned these. The announcements came across repeatedly on the radio and conveyed both urgency and terror: "Tsunami warning—Please evacuate!" "Fire—Please evacuate!" Not one moment to spare. Furthermore, the city's annual drills made all the difference. Recovery workers wasted no time in putting up a giant tent in the parking lot of the destroyed city hall. At 3:28 p.m., the mayor issued mandatory evacuation and commissioned staff to examine the extent of damage in Tono.

The Tono Sports Park became the logistic headquarters for city officials, members of the Fire Department, police, and the SDF. These headquarters later proved to be a vital center to coordinate aid from other cities like Shizuoka over 700 kilometers (434 miles) south of Tono, and Tokyo, 554 kilometers (344 miles) also to the south. Indeed, all cities in Japan sent relief, and sent it through Tono. Ogushi, Ishin, and Kikuchi sent employees to Tono headquarters. Many cities and companies used Tono as the headquarters for relief aid. Besides citizens, medical professionals, police, firemen, and SDF, another source of manpower were volunteers, like the later formed Tono Magokoro Net, the group I worked with in winter 2014. The term *magokoro* means "true heart," and nicely captures the spirit of the people of Tono.[28] (I will describe my volunteer experience in a later chapter). See figure 3.1 for a photo of Magokoro Net headquarters in Tono.

Tono has around 30,000 residents. Yet volunteers who came to the city eventually numbered close to 100,000, filling the small town far beyond capacity. Nonetheless, the city was able to accommodate them by opening up 134 facilities in schools, community centers, and some regional health centers. There were absolutely no complaints from citizens concerning this large number. Organizations continue to use Tono as a base to communicate with each other and coordinate relief efforts.

"There are 500 people seeking refuge at the Otsuchi High School. They have no food or water and need your help."[29] Nearly twelve hours after impact, on March 12, at approximately 1:40 a.m, a survivor from the coastal disaster reached Tono on foot. After walking nearly 44 kilometers (27 miles) through ruined roads, climbing steep paths, and crossing two mountains, each around 1,000 meters (3,200 feet) high, a young man in his twenties from Otsuchi-cho appeared before city officials and reported terrible devastation in Otsuchi and in neighboring Kamaishi. The young man lives in Kamaishi. His wife, who was working in Otsuchi High School, survived the waves. The

Figure 3.1. Tono Magokoro Net headquarters, 2014

young man managed to cross the steep mountain terrain to alert Tono. He pleaded "We must help." Within hours, citizens and officials gathered together and set out for Otsuchi at around 4:50 a.m. Fire department members led the way. Many other citizens, risking their safety and facing all sorts of hazards, climbed down the steep and treacherous mountain paths to help their stricken neighbors. They carried with them life-saving supplies like water, rice, clothes, blankets, boots, and umbrellas. Umbrellas were crucial during this rainy and snowy season. From that moment on, Tono became the lifeline for victims and survivors in the crushed coastal communities.

To claim that Tono citizens all worked together in helping victims and survivors remains an understatement. Everyone in Tono played a role. This was a city-wide effort. All able citizens played a vital part in sending whatever aid they could muster to their neighbors on the coast. Women cooked countless bowls of rice, making over 120,000 rice balls. For weeks and months afterwards, citizens everywhere gave away their clothing, umbrellas, and blankets from their own homes. High school students helped pack countless boxes with supplies, food, and clothing. Simply put, assistance to victims and survivors could not have occurred without the citizens of Tono. Once some roads were cleared from the earthquake's debris, Tono officials and citizens sent busses with food. Even coastal libraries left standing were attended to. Junior high school students collected monies to help restore one of the local libraries on the coast and helped in cleaning its books. Elementary school students also collected monies for relief aid. They also worked together in putting on a performance to collect aid. The performance was held in Tokyo's major theater, attended by the royal family. Evacuees from throughout the coast later came to seek temporary shelter in Tono. Tono citizens offered up their homes to survivors. In some cases, around thirty people would stay in one house for over three months. There was strong people-to-people connection and durable ties among Tohoku people. Survivors were given food and food coupons. They came to Tono with nothing, perhaps one candle, one drop of fuel, one rice ball. No money. Citizens of Tono as well as evacuees who came to Tono purchased food and water via invoices. And when power was down for over a week in Tono so that cash registers were shut down, they could pay later through a built-in system of trust. Trust, the most collective of all virtues and the oxygen of human interaction, was all-important.

Not only had 3.11 knocked out all power lines, but there was little fuel. All along the coast, there were decaying dead bodies. There was no fuel for cremating the bodies. What could be done with these bodies that would still respect the integrity of victims? When surviving mayors were able to locate and speak with many family survivors, they suggested that bodies be buried temporarily. There was no other choice. Eventually, some of the dead bodies were transported to Tono for cremation. However, there was a snag. Tono had just one cremation facility (*kasōba*). Although the crematorium daily holds around two cremations, it could accept at most only six bodies per day. Why only six bodies? Practically, there was little available fuel. Also, cremation in Japan is an elaborate process that requires the proper amount of time. Typically, the entire process takes anywhere from 90 minutes to two hours. The facility was also facing logistical issues in storing and handling so many dead bodies. This meant that cremations needed to be fairly allocated. In view of the customary six bodies per day, city officials decided to allot four cremations per day to disaster victims, and the remaining two per day for citizens. Priority was given without question to 3.11 victims.

MAYOR TOSHIAKI HONDA: GRIT AND *GIRI*

With his firsthand knowledge of the Sanriku coastal topography and former role as the region's Deputy of Disaster Preparedness, Toshiaki Honda was the right person at the right time to lead his town. During the 1995 Hanshin Earthquake in Kobe, Honda was an Iwate Prefectural employee assigned to the disaster management office. In 2002, he was elected mayor of Tono. The

Hanshin Earthquake was at that time the largest earthquake in post-war Japan. Since then, the national government required close studies of Japan's geographical fault lines. As a prefectural employee, Honda carefully studied Iwate's disaster history and its topography. He meticulously examined Iwate's coastal towns and cities and even walked along these coastal sites. In the process, he examined effects from the 1896 Meiji-Sanriku earthquake and the 1960 Great Chilean earthquake, also named the Valdivia Earthquake. Along the coast were memorials and signs that marked the heights of tsunami waves from 1896 and 1960. This experience furnished an invaluable opportunity to study the effectiveness of helicopters during evacuation as well as the entire evacuation process in view of an area's geography and topography.

Here we see the literal significance of "climate." Through his detailed, in-depth, on-the-ground study of the Sanriku coastline, its complex configuring of bays, inlets, hidden harbors, open-faced coastline, layers of land, areas of high ground, roads, accessibility to high ground, locale of surrounding hills and their steepness, when 3.11 struck, Honda had a much clearer image of the possible impact on the coastal towns. Honda also admits that without this prior knowledge and visits to the coastal areas, he may well have waited, as officially required, for prefectural government directions instead of breaking rank with law and taking action on his own, which he did.

This prior knowledge and experience enabled the mayor, soon after he was elected, to put together a plan to use Tono as a base for operations in the case of a major disaster. Under the supervision of his dauntingly titled Sanriku Region Earthquake Disaster Logistic Support Base Provision Implementation Council, which he established in 2007, Tono was the operational base for coordinating aid from surrounding municipalities, SDF forces, and fire, police, and medical support systems during disasters. This headquarter practiced regular drills, with full participation of SDF, police, and fire departments. From past experience, it was logistically clear that the Tono Sports Park would be the proper site for disaster headquarters. In fact, it hosted the largest drill of its kind in 2008 through the Northeastern Army Disaster Response Training program on October 31 and November 1. It simulated response to an earthquake with its epicenter off the Miyagi coast, followed by a tsunami sweeping the Sanriku coastline. The drill involved 18,000 people, 43 aircraft, and over 2,000 vehicles.[30]

All this eventually paid off. The prospect of an impending disaster was becoming apparent. In 2007 and 2008, experts predicted a 90 percent likelihood of a major earthquake striking the Miyagi region within the next thirty years. Given the earth's long history, thirty years is like tomorrow. The mayor himself confessed to me that he had a personal premonition that the earthquake would happen well before the end of thirty years.[31]

Beyond Rules

Japan is a formidable rule-oriented society. Laws play a crucial, indeed domineering role, and there is immediate deference to the authority of law and official protocol. I know this first-hand. When I conducted research in Tokyo years ago on the issues surrounding Japanese views regarding heart transplants, I especially found this to be the case. In order to conduct official interviews I needed to go through the customary channel of formal requests, introductions, and invitations. The process is often cumbersome and time-consuming.[32] Rules also played a pivotal role for us while we were volunteers with Tono Magokoro Net. There were regular allusions to rules at the facility, rules about cleaning, keeping the kitchen in order, lights out, boarding vans in time for departure to the coast, and so forth. One group leader, Sudo from Hokkaido, once apologized to me for having to throw out my yogurt, milk, and orange juice that I had stored for three days in the community refrigerator. Though the contents were fine, food items could only be left in the refrigerator for no more than two days.

Here, the Tono mayor stands out. In emergencies, long-standing laws presume that mayors of affected towns must first officially request help from the prefectural and national government before moving forward. Official and legal protocol as specified in the Disaster Relief Act requires that local fire department and police, working with SDF, would respond accordingly only after the initial contact and request.[33] But in the case of March 11, due to the scope of damage, spreading over 500 kilometers (310 miles) from Tohoku to Tokyo, it was practically impossible to take this first step and ask for help. So, convinced that he and his citizens were *directly* responsible for saving the lives of their neighbors on the coast, the mayor took immediate action despite the existing laws. When the quake violently jolted his city and his office trembled so uncontrollably he and his staff could not stand, the mayor's immediate thoughts were: "Are the citizens and city officials safe? Is there fire? What about the bridges. Are they collapsing? What about the nursery schools? Elementary schools?"[34] He also knew that with such violent trembling a tsunami would certainly follow in its wake and deliver terrible damage on the coast. He knew that the extent of damage and casualties from a tsunami depends very much on geographical, topographical, and demographic features of the area. Herein lies our inescapable *aida*, in-betweenness, our unavoidable situatedness within our climate, weather, geography, and topography. The mayor was well acquainted with the coastal topography, and realized that because the coastal cities are near each other, every city, town, and village along the coast would be dangerously impacted. The devastation would be extensive, with numerous casualties. The mayor donned his disaster relief uniform, and then and there, at his heavily damaged office building, he was committed to work from there and his later makeshift office for the next week, even though his intact home was only ten minutes away.

The dreaded aftershocks came as expected. Most citizens including the mayor's wife began to evacuate to the city's shelter. In the early morning on March 12, the young survivor from Kamaishi reached Tono. His desperate plea for help reached Mayor Honda. And from that moment on, the city's leader worked around-the-clock, unflinchingly, without sleep or rest. City officials followed the example of the charismatic mayor and worked stubbornly nonstop as well. So did the city employees and citizens throughout the city. They simply had to. There was no choice. The coastal victims were their neighbors, and they needed help. Bear in mind that Tono's relationship to the coastal areas and its inhabitants ran deep in Tono's history, with families living there, and strong connections through marriages. For example, many Tono citizens had worked at Kamaishi's huge iron factory and had a long history of relationships. Although Kamaishi is 40 kilometers (over 24 miles) from Tono separated by 1,000 meter (3,200 feet) high mountains, Tono citizens eventually supported over 100,000 Kamaishi victims by providing food, clothing, water, and shelter.

Tono citizens and officials all knew that immediate response was vital. The term "response" is instructive. It does not simply denote acting, but acting with responsibility. Manpower was urgently needed to go into the afflicted areas along the coast, search for survivors, and provide aid in the quickest ways possible. Under orders from their mayor, Tono city officials responded to the crisis. The national and prefectural governments were not functioning sufficiently for nearly a week after 3.11. What did function were the local municipalities. They and the local citizens were in effect the front line of response.

When the World Trade Center towers crumbled on September 11, 2001, the heroic actions of the New York City Fire Department and Police Department leave their indelible legacy. The other "heroes" that day, less recognized, were ordinary citizens, unsupervised by authorities, doing whatever they could to assist the thousands who were injured, exhausted, and confused. They formed spontaneous groups providing water, food, clothes, supplies, and shelter. There are numerous accounts among the over 20,000 employees in the towers who managed to escape while aiding each other in orderly fashion, from coworkers carrying their paraplegic colleague down sixty-nine flights of stairs, to overweight people stepping aside on the way down so as to not disturb the flow of evacuees. Despite differences, all had common cause in mutual survival, like the Hasidic Jewish man who aided young Pakistani immigrant Usman Farman in running away from the sky-high, looming infernos. Rebecca Solnit describes these and other inspiring accounts in her engaging *A Paradise Built in Hell*.[35] She recounts all varieties of altruism on the part of ordinary citizens who acted instinctively as first responders to the calamity. Like the two strangers who stayed by the blind Maria Georgina Lopez Zambrano who ran a nearby newsstand. And the armada of boats who came to the scene from New Jersey, Staten Island, Long Island, and elsewhere to help evacuate hundreds of thousands of people.[36] The true heroes that day were ordinary folks, perhaps a natural result of close-knit urban living. Solnit suspects the same would not be the case in the suburbs. She writes:

> New Yorkers were well served by their everyday practices of walking the city, mingling with strangers, and feeling at home in public. It is hard to imagine many of the more suburbanized and privatized American cities responding with such resilience, resourcefulness, and public-spiritedness, and so the everyday qualities of true urbanism may too be survival skills in crisis.[37]

Norway is historically bound in the value of trust. Given its past homogenous society, cultivating trust comes more easily. As Are Holen stressed when we met in Oslo, the Norwegian character is one that cherishes trust, comraderie, solidarity, respect, and equality. This is why the July 22, 2011, Utøya camp massacre came as a collective shock. People asked, "How could this happen here?" When the deadly explosion rattled government offices in Oslo and a lone gunman killed young campers point bank at nearby Utøya island, ordinary citizens and neighbors were the first on the scene. Neighbors overlooking the island risked their safety by immediately boating to the island after hearing gunshots to save many young campers. Most campers were saved due to these neighbors' efforts. Afterwards, during the investigation into the disaster, Oslo police came under fire for their slow response.

In a perfect world, laws are moral. Our world is far from perfect. We therefore face the classic perennial tension, not always a healthy one, between doing what we think is moral and doing what is legal. For someone like Toshiaki Honda, a man of moral conscience and fully committed to serve his people, it was crystal clear what needed to be done. Lives were at stake and time was of the essence. Legally, however, he was handcuffed. But there was simply no time for waiting. The mayor knew that he had to act immediately and send whatever aid he could muster to the helpless victims on the coast. Here again lies the consequence of *aida*, our inescapable in-between. Precisely *because* Tono was not a coastal city, yet conveniently situated near the coast, the mayor felt that his people had a moral imperative to help the coast 100 percent. His reasoning was straightforward. Tono citizens are not tsunami victims. But the people of Tono are neighbors, in-between the coastline and the mountainous highlands. Their situatedness compelled immediate action. Many of us may customarily think the reverse: "Since we are far enough away from the tragedy, it happened 'over there' and 'to those others.' Therefore we are less connected, and less obliged to help." Yet *exactly because Tono citizens are not coastal, and therefore not victimized, they are all the more connected to the victims* and thus have an immediate duty to help the victims. The mayor's reasoning embodies the power of *aida* in his moral judgment. His judgment is the basis for moral grit.

Mayor Honda confessed to me a number of times that he felt as if he was acting "like a dictator (*dokusaisha*)" in the small town. *Dokusaisha* is too strong a word. The mayor, actually rather humble in demeanor, certainly does not come across as a dictator. What he means is that he had to be strong and exceptionally assertive. Ultra-assertiveness is uncharacteristic of Japanese leaders. For weeks following 3.11, he ordered his employees to come in to work from six in the morning through eight at night. He ordered all employees driving to work to do so in car pools in order to ration the city's scarce supply of gasoline. In addition, because the fiscal year ends in March, reassignment of municipal posts generally occurs that same month. However, the mayor froze all reassignment orders. For executive officials about to retire that month, their retirement was put on hold. All other shuffling of positions had to wait. The mayor knew the risks of his breaking rank. Should adverse events occur, like landslides (typical in the aftermath of earthquakes) or injury to workers and citizens, he would have to bear the responsibility and even be forced to resign. His citizens were risking their lives climbing down and up dangerous mountain paths and entering flooded areas thick with hazardous debris. Needless to say, he regularly expressed relief when recovery workers returned safely from the stricken coastal villages.

Moral Grit

Here is moral grit in action. The mayor acted beyond the standing laws. His first duty, *giri*, lies in ensuring his own citizens' safety. His second duty is to ensure the safety of others. Without a second thought, he was morally compelled to send help to victims and survivors, relieve their suffering and save lives in whatever ways possible. He demonstrated defiance through his positive spirit and willpower, and embodied hope all throughout. He instructed his citizens and recovery workers to give everything possible to evacuees. These evacuees have lost everything but their lives. And through his example, citizens together displayed their heartfelt moral imperative to relieve others' suffering.

The mayor is now working to better prepare his city for future disasters. He is establishing a new anti-disaster education center within the city office. The city office is still in its temporary quarters. That's where I met with him. He hopes to have the new center under way once the original city office is repaired. Whatever measures officials take, the key lies in never forgetting 3.11. Citizens must continue to share with others what happened that day and afterwards. History will always be our most powerful teacher. So will necessity. Circumstance often forces us to recognize our inescapable in-between. As Mayor Honda puts it, "the 3.11 earthquake forced us out of necessity to recognize our ties to the coast, and share this with the world."[38]

When the wave smashed into Otsuchi, an older couple, in their seventies, immediately left their home. Though they lived on higher ground, upon seeing the enormous size of the oncoming wave, they knew they had to head quickly to the nearest evacuation shelter. Stumbling along the way, they made it safely to the designated shelter. The husband then remembered their next door neighbor, an elderly lady in her eighties who lived alone. He went back as quickly as he could and entered her home. He then carried the lady out on his back, staggering back to the shelter. But there was not enough time. His wife spotted them from a distance in the shelter as the giant black wave came and swept them away.

The mayor related to me this story. This made a lasting imprint on his memory. The widow was one of the evacuees who later sought shelter in Tono. When the mayor met with her, she did not express grief. Instead, she stated "I'm proud of my husband. My husband was great." And she repeated, "Mr. Mayor, I am so proud of my husband." Her remarks left the mayor speechless. He tells me, "I did not know how to respond. I could not say to her, 'I'm so sorry.'"

There are many commentaries and articles about the Japanese sense of orderliness during the disaster. There were no riots. No stealing. Many people overseas, particularly Americans, expressed shock at this almost surreal reaction. When I mentioned this in speaking with the mayor, he pointed out that this is not unusual and has to do with the character of the Tohoku people. "I can honestly say that many in Tohoku region know how to live with dignity, live in the situation and withhold emotional response." Throughout their history, the people of Tohoku have regularly faced the challenges of their environment, harsh climate, and weather. And facing this challenge together has further fostered a sense of *gaman*: courage, and inner strength. 3.11 was a litmus test of this. The wife above, so proud of her husband, shows this.

"There is no recovery unless everyone is recovered." Akiko Iwasaki, my innkeeper friend in Kamaishi, whom we will meet in the next chapter, repeatedly stressed this to me. A few months after 3.11, in July and August, the national government planned to move many survivors from the stricken coastline further inland to have them live in available hotels. The evacuation shelters were overcrowded. However, to the surprise of officials, many of them refused to go. Why? These hotels were surely more comfortable than living conditions in the packed evacuation centers. Yet, for these evacuees, their reasoning was less about comfort and more about fairness. Each responded, "Why only me? There are so many other evacuees who have lost family, homes. They also suffer. Why just me?" As Mayor Honda emphasizes, "Tohoku people are not selfish. This is the natural way of thinking for Tohoku people." As he puts it, they have this inner strength, *gaman*. It is built into the "DNA of Tohoku people."

As of this writing, well over three years since the disaster, the mayor feels dismayed that reconstruction is proceeding at snail's pace. One major reason has to do with establishing proper land ownership. Land ownership can reach far back in generations. Authorities must first obtain consent from the legal owners, who are usually related, but who could be living anywhere throughout Japan. The issue is exceptionally complex. Concerning this issue, here is a summary, or paraphrase, of a brief part of my conversation with Mayor Honda.

> Toshiaki Honda (TH): The government is moving forward according to existing laws and systems, which in themselves cannot cover all contingencies. An example of this concerns ownership of land and property. Local city governments and prefectures want to build houses further inland. But, because of questions surrounding proper land ownership, the national government opposes any new building since its Constitution guarantees permanent rights related to land ownership.
>
> MB (author): Does land belong to individuals who buy the property, such as a farmer? Does he own the land? Or does he just own the facility or building?
>
> TH: The land is privately owned, so that the national government prohibits building on and moving onto privately owned land. Local government cannot interfere in peoples' property rights. And here is the problem. The Constitution also guarantees the right to a decent life. From my point of view, people's right to this decent life takes precedence over property rights. But there is little discussion about this among officials.
>
> MB: Are private owners of land willing to sell or give their land to people from coastal regions?
>
> TH: Many land owners are certainly willing to sell their land. But, the inheritance of their land and land ownership rights reach back many, many years and many generations. The issue is very complicated. Currently, mayors of coastal towns are planning to build housing on the edges of their cities. But they need to first figure out who is the land owner. In many cases there are multiple land owners. So, city officials are now visiting the rightful land owners from Hokkaido [in the north] to Okinawa [far south], meeting with them personally, talking with them, and hopefully getting their signed consent and signatures.[39]

This quandary compels us to ask: Will the afflicted communities ever recover? According to the mayor, much rests upon the strength of these communities and their social cohesiveness. This is enormously challenging in a Japan that has become increasingly individualistic. A journalist from an international newspaper interviewed Mayor Honda shortly after 3.11. She asked, "Why are the people from Tono helping these coastal cities and their survivors. Is there something they hope to gain from helping them?" The mayor, shocked at her question, shared with her Tohoku's long-standing respect for tradition, ritual, and community and how these formed a basis for the town's actions. The Tohoku region still holds fast to its traditional values and sense of community. This solidarity will enable it to keep hope alive, embody moral grit, and recover. But given the magnitude of the disaster and its scope of destruction, this will also take time. Lots of it.

The mayor believes there are invaluable lessons to be learned from the calamity, particularly amidst a climate of growing individualism and isolation. He states that "the earthquake was telling us that human beings cannot survive alone. We need human connections, community. That's what the disaster tells us." This may help explain why marriages have been on the rise since the disaster. Many seek a committed relationship. The disaster offers deep lessons for interpersonal connecting—connectedness, not connectivity. This connectedness entails the dynamic vibrancy of face-to-face, interpersonal encounter and engagement.

But there's a deeper lesson. This has to do with *how* we communicate with each other. As the mayor states, "the disaster is also teaching us the importance of communication." He is referring here to connectedness, not connectivity. Connectivity via devices means "communicating" without actually speaking to a person, though you can "speak" by way of a device or screen. And, without doubt, this happens now throughout the world. Even in Tono, as Mayor Honda admits, it's "possible for people to go through the day without speaking with anyone else. Because everything now is mechanized, we push buttons, tickets come out. There is no need to speak with a station employee. When we want food, we can just purchase the food without saying anything." Convenience stores have become a major part of Japan's town and city landscape. And even here, we can simply put food into the basket, proceed to the register, and pay without speaking. But it's not the same.

NOTES

1. In their stirring account of the events and aftermath of 3.11, Tokyo-based reporters Lucy Bimingham and David McNeill describe this eerie incident. Lucy Birmingham and David McNeil, *Strong in the Rain: Surviving Japan's Earthquake, Tsunami, and Fukushima Nuclear Disaster* (New York: Palgrave Macmillan, 2012), 153.

2. Kunio Yanagita, *The Legends of Tono*, 100th Anniversary edition, trans. Ronald A. Morse (Lanham, MD: Lexington Books, Rowman & Littlefield, 2008; orig. *Tono monogatari*, 1910), 37.

3. Yanagita, *Legends of Tono*, 35.

4. Birmingham and McNeil, *Strong in the Rain*, 176.

5. Alan MacFarlane, "The three major famines of Japanese history," accessed Dec. 20, 2014, http://www.alanmacfarlane.com/savage/A-JAPFAM.PDF.

6. In George Sansom, *A History of Japan: 1615–1867* (Stanford, CA: Stanford University Press, 1963), 185; cited in MacFarlane, "Three major famines," 2.

7. Jean-Pascal Bassino, "Market Integration and Famines in Early Citizens Modern Japan, 1717–1857," accessed Dec. 21, 2014, http://federation.ens.fr/ydepot/semin/texte0708/BAS2007MAR.pdf.

8. M. William Steele, "The Great Northern Famine of 1905-1906: Two Sides of International Aid," accessed Dec. 20, 2014, http://subsite.icu.ac.jp/iacs/PDF/ACS39_PDF/ACS39_01Steele.pdf.

9. Steele, "Great Northern Famine," 2.

10. Steele, "Great Northern Famine," note 10, 3.

11. Steele, "Great Northern Famine," note 37, 11.

12. Steele, "Great Northern Famine," note 44, 12.

13. "*Shinsai kara hantoshi Tōhoku yo–'Tōhoku gaku' o teisō suru minzoku no gakusha. Akasaka Norio-san*" [Six Months Since Disaster in Tohoku: Folklorist Akasaka Norio Advocates "Tohokuology"] *Asahi Shimbun* (Sept. 10, 2011): 15, in Anne Allison, *Precarious Japan* (Durham, NC: Duke University Press, 2013), 201.

14. Nathan Hopson, "Systems of Irresponsibility and Japan's Internal Colony," *The Asia-Pacific Journal* 11, issue 52, no. 2 (December 30, 2013), accessed Feb. 6, 2015, http://japanfocus.org/-Nathan-Hopson/4053.

15. Katsurō Hara, *An Introduction to the History of Japan* (New York: G.P. Putnam's Sons, 1920): 26–27, cited in Hopson, "Systems of Irresponsibility," http://japanfocus.org/-Nathan-Hopson/4053.

16. Yuzuru Tsuboi, "Interview: Anthropologist say Tohoku region must gain independence from Tokyo," *Asahi Shimbun*, August 1, 2014, accessed Feb. 6, 2015, http://ajw.asahi.com/article/views/opinion/AJ201408010006.

17. Tsuboi, "Interview," http://ajw.asahi.com/article/views/opinion/AJ201408010006.

18. Watsuji Tetsuro, *Fudo: Climate and Culture: A Philosophical Study*, trans. Geoffrey Bownas (Westport CT: Greenwodd Press, 1988; reprint, orig. publisher Tokyo: Monbusho, 1961), 6, 8.

19. Noted Waseda University philosopher Tomoaki Tsuchida stressed this diversity in my conversation with him, March 25, 2014.

20. Interview with expert in psychosocial medicine Makiko Okuyama, March 24, 2014.

21. There's an interesting inconsistency here. In Tono, though it desires to increase tourism after 3.11, most restaurants do not accept credit cards, and it is difficult for foreigners to obtain cash at local ATMs. On the other hand, nearly all toilets are modern and heated. The warmest spot in our Spartan volunteer headquarters where we relied on kerosene heat were toilets with their heated seats.

22. Interview with Mayor Toshiaki Honda, March 5, 2014.

23. *Tōno Monogatari* was made into an award-winning animated film in 2007 directed by Kenichi Hara, "Summer Days with the Kappa Coo." It is also the basis for a manga (comic book, extremely popular in Japan) magazine series in "Big Comic" rendered by the artist Shigeru Mizuki. Mizuki is known for his images of ghosts and goblins.

24. Masatsugu Orui, Shuichiro Harada, and Mizuho Hayashi, "Changes in suicide rates in disaster-stricken areas following the Great East Japan Earthquake and their effect on economic factors: an ecological study," *Environmental Health and Preventive Medicine* 19, Issue 6 (November 2014): 459.

25. To learn more about Furosato Village, see http://www.tono-furusato.jp.

26. The website for Tonojikan is http://www.tonojikan.jp/Several_languages/english/english.html.

27. Hiroko tells me that they are strict in teaching the children there. Tourists sometimes view this as bordering on abuse.

28. From an email correspondence with Makoto Yanagisawa, Magokoro Net's Public Affairs Manager, March 12, 2015.

29. In Report by Tono City Council and Mayor Toshiaki Honda, "Logistical Support in Tono's Coastal Disaster Zone: Community Ties Aimed at Reconstruction through Relationships," March 5, 2014.

30. Tono City Council, "Logistical Support," 4.

31. Interview with Mayor Toshiaki Honda, March 5, 2014.

32. It helps to have an intermediary who is Japanese. It also helps to be born in Japan of a Japanese mother. For my visits to Tohoku and interviews with the mayors, Kunio Aoki and Hiroko Suzuki were my intermediaries. For my research on heart transplants, the University of Tokyo's Akira Akabayashi acted as my contact. Akira, one of Japan's leading bioethics scholars, directs the university Center in Biomedical Ethics and Law. They were all so gracious, kind, and invaluable.

33. Tono City Council, "Logistical Support," 9, citing Toyu University Research Center Report No. 15.

34. Interview with Mayor Honda, March 5, 2014.

35. Rebecca Solnit, *A Paradise Built in Hell: The Extraordinary Communities That Arise in Disaster* (New York: Penguin Books, 2009), 187–88.

36. Solnit, *Paradise Built in Hell*, 189, 191.

37. Solnit, *Paradise Built in Hell*, 188–89.

38. Interview with Mayor Honda, March 5, 2014.

39. Interview with Mayor Honda, March 5, 2014.

Chapter Four

Akiko's Lantern

Our journey from Tono City to Kamaishi took an entire day. This was in January 2012, nearly ten months after the calamity. Kunio Aoki who manages East Japan Railways's Culture Foundation rented a car in Tono and the four of us—Kunio, his assistant Mizuki, Hiroko, and I—were on our way to meet Akiko Iwasaki, an innkeeper in Kamaishi. Though three hours by car, we stopped along the way, stunned, overwhelmed, and speechless. The spectacle was unrelenting. We drove through beads of entirely flattened towns. The sites were painfully similar: leveled cities, towns, and villages; hollow buildings; gutted skeletons of the few homes left standing; collapsed concrete and steel bridges; capsized cars and boats strewn chaotically among the remains; toppled high-walled levees intended to hold back an angry sea; endless miles of giant mountains of debris assorted according to wood, plaster, concrete, metal, and paper; and surrounding farmlands buried under sand. On that ill-fated March 11th, the sea barreled inland as far as six miles. During the calamity, blazing fires from fuel-filled, colliding, and overturned vehicles devoured villagers. Blazes left scarred pockets throughout the landscape. Thousands of living beings were burnt to a crisp before monster waves spat out the flames. As we drove from one graveyard to the next, screams of death still lingered in the air in utter silence.

Kamaishi was at one point home to Japan's steel industry. Nippon Steel was its biggest employer. In the 1970s, Nippon Steel moved to central Japan to support its auto industry renaissance. With the move of Nippon Steel, who employed most of the town's 40,000 residents, Kamaishi's economic future was at stake. The town put on a full-court press to lure more fishing and maritime-related industries to Kamaishi. Kamaishi eventually became a thriving center for tourism and commercial fishing, known for its salmon and scallops. To ensure protection for its people, fishing, and future, the town invested enormous sums in a new seawall. This proved to be folly on March 11. Waves easily crashed over it and demolished much of the city, taking 1,046 lives out of a population of over 39,000. Nearly 160 remain missing.[1] By the time we reached the small coastal hamlet of Kamaishi and its ryokan, or traditional Japanese inn, the four of us were emotionally spent. After our trek through dead towns, Akiko and her gentle staff's warm welcome blessed us with a much-needed medicine.

OUR RYOKAN

Akiko's ryokan, named "Houraikan," as with most traditional inns, incarnates a good bit of Japanese culture and custom. Spending time in a ryokan remains one true way to get a sense of Japan's traditional lifestyle, which, as we saw in Tono City, is still very much alive in bustling, contemporary Japan. With its gardens, stone lanterns, hanging scrolls, flower vases, and authenticity of presence we found our haven of serenity, particularly after our heartrending crossing. Here was a chance to reconnect with simplicity and tranquility, what Japanese call *wabi.* The ryokan's link with Nature is so apparent. Like a traditional Japanese house, Houraikan's openness, tatami flooring, alcoves having scrolls, *ikebana*, or *bonsai* (these alcoves are called *tokonoma*), and sliding paper panels, or *shoji*, allowing plenty of light, ventilation, and views of surrounding gardens and nature.

Originating as far back as the Nara Period (710–784), these inns were rest stops, *fuseya*, for weary travelers, often run by Buddhist monks. As pilgrimages to Shinto and Buddhist shrines increased, these developed into lodgings offering meals. With improved roads, these *hatago* inns became popular during the Edo Period (1603–1867). The more elegant inns, *honjin*, were reserved for feudal lords (*daimyo*) and their entourage. Travelers also visited hot springs for health treatments. Hot springs lodgings, *tōji-ba*, developed into hot-spring ryokan, or *onsen ryokan.* Visits to *onsen* are a long-standing practice, dating back over one thousand years and referred to in the classic *Chronicles of Japan* (*Nihon Shoki*). Here is a mixed blessing in this land of volcanoes. Along with the threat of earthquakes and volcanic eruption, hot springs and their soothing geothermal heat have been an enduring source of serenity and healing. Japanese faithfully visit these hot springs from all over the country. Eventually, with railway systems enabling better ease of travel, ryokans were often built near the railway stations. The ryokan now combines the practicality of the traditional *hatago* with the graceful yet simple refinement of the *honjin*, and its essence lies in offering hospitality to travelers and pilgrims. For me, this refuge with its core Japanese values was a much-needed pause on our journey. This pause reminded us of our natural in-betweenness, the intimate bond we share with nature and with each other.

The ryokan also reflects the seasons. We arrived on January 8, just after *Shōgatsu*, those first three days of the New Year when, remarkably in Japan, no one works. January is called *Mutsuki* (month of relations of harmony), and its New Year celebration stretches from January 1 through 7, a period called *Matsu-no-uchi*. Our ryokan displayed New Year's symbols like decorated pine, or *matsu-kazari*, and Shinto's sacred twisted rope, *shime-nawa*. The pine came from tall, thin pine trees covering the hill behind the ryokan. This was the same hill that offered sanctuary from monster waves to villagers and the ryokan's escaping guests on 3.11. In these ways, the ryokan reminds us that we are creatures of our seasons and climate, and we carve out our worldviews, values, and beliefs according to our place in this natural and human climate. As Watsuji Tetsuro emphasizes, we are *aidagara*, "in-between." We are innately in-between, imbedded within the change of seasons, natural and human. Natural and human intimately intertwine. One truth remains certain in our synergetic interdependence with nature and with others—that of Nature's pervasive, titanic power. Nature is our ultimate teacher, implanting in us meanings of life and death.

Enduring its natural calamities of earthquakes, typhoons, floods, and snowstorms, Japan is no stranger to Nature's rhythms, especially since the sea surrounds its islands. There's no better way to demonstrate nature's interconnectedness than through the sea. Note the ongoing conversation among the atmosphere, the ocean water, cloud density, and the winds. Sailors and fishermen intuit the natural tension between wind currents and tides. To better understand the sea they live and work on, they learn from experience how to read the skies. And what takes place at sea indelibly affects land. All of nature is situated in-between all phenomena. We humans participate in this rhythm, and our footprint is both effect and cause. Our in-between (*aida*) is inescapable. And one dominant truth we learn from our in-betweenness is that nothing stays the same. All life is change. Today is not just another day. Neither is tomorrow, and that goes for every tomorrow.

We often learn of our unavoidable in-betweenness the hard way. Our love of the sea lures us into buying our beachfront cottages by the ocean. But when that hurricane hits, with its powerful winds, rapid rise in sea level, and unwanted invasion of saltwater, our climate becomes more real. Our situatedness cries out to us. All this instills a natural awe and wonder at Nature's power. We can never discount or quiet the intractable voice of Nature, pleading "Here I am." And according to Japanese long-standing indigenous Shinto beliefs, sacred forces called *kami* permeate the natural order, gods and goddesses of wind, sea, rain, earth, rice-fields, and so forth. Seasons arrive with their special blessings, and Japanese meet these seasonal blessings with gratitude as well as awe. It all comes down to living with and adapting to Nature rather than attempting to control it. Put simply, we humans are not the measure of all things. Nature is. We belong to Nature. Nature and human are intrinsically linked, and Japanese honor this intimate bond. This is also the bond we have with each other, for the other is also part of Nature, and part of us. We thus remain creatures "in-between."

Back to Houraikan. That evening, we immersed ourselves in the inn's *daiyokujo*, or large common bathtub. The inn had separate common baths for men and women. *Daiyokujo* helped cleanse my physical and emotional fatigue. When Buddhism entered Japan in the sixth century, it introduced the practice of bathing as a virtue. Bathing became a way to purge away anything polluting like dirt, dust, and perhaps most polluting of all, death. In ryokans, bathing offers the traveler an opportunity to be free from the journey's stress and fatigue. The *daiyokujo* also reflects our in-betweenness, a way to relax body and mind together to share in the hot water's catharsis. It was for me a touch of heaven, as was my night's sleep wrapped in the most comfortable down-feathered futon.

Akiko

In his magical tale, "The Nightingale," Hans Christian Andersen tells us that one day the emperor of Japan learned that the emperor of China had a special guest in his porcelain palace. This special guest was the famed nightingale who inhabited the forest and was known throughout for her beautiful music. The Japanese emperor then sent the Chinese emperor a gift. It was a splendid, mechanical wind-up bird decorated with precious gems. This was a marvelous instrument. Simply winding the key to the bird enabled the artificial nightingale to sing a programmed pattern of songs. The Japanese emperor attached a small note to his gift: "The Emperor of Japan's nightingale is poor compared with the Emperor of China's."

What did he mean? Why was the Chinese emperor's nightingale more precious than the wind-up bird? Because the nightingale's song was real. She sang songs of truth, of "those who are happy, and those who suffer; of the good and the evil hidden around you," of joys and sorrow, of loss and despair, and of discovery, promise, and hope. In contrast, the wind-up bird's song was planned and expected. Its songs removed contingency. And contingency, what is unforeseen and unplanned, defines life.[2]

Akiko Iwasaki is Houraikan's owner and manager, or *okami*. As the inn's *okami*, she bears the responsibility of ensuring the comfort of her guests, satisfaction of her staff, and sustaining close cooperation with the local community, all of which she does with grace and genuineness.[3] Akiko soon introduced us to her daughter and granddaughter who have stayed with her since the disaster. Someday, her daughter will likely take over her mother's professional role at the inn. Like her mother, she wears a kind face, trying hard to hide her own inner sorrow for having lost her best friend and high school classmate. On March 11, her friend lost her entire family to the sea. And she had been missing for weeks. Akiko's daughter strongly suspects she took her life, probably throwing herself into the cold, deep waters off the nearby rocks. Others have done the same along that coastal region, those who lost everything—family, home, property, friends. Though these suicides do not make headline news, they remain a public secret.

The hospitality of Akiko's staff was palpable. Room maids, *nakai-san,* dressed in traditional kimono, were serenely gracious. Keiko, head of the kitchen staff, particularly displayed this grace. Tall, heavy-set, and maternal in bearing, Keiko made sure we were satisfied with our meal and cared for properly. Keiko has soft yet sad eyes. She had lost her husband and son to the waves. There's a special phrase in Japanese, *ichigo-ichie*, which nicely captures the hospitality of Keiko, Akiko, and other staff at the inn. It also sums up the quality of my encounters with Hiroko, Tono residents, and many others. This phrase first came into use during *Sado*, the traditional tea ceremony, in which each movement and moment is conducted in complete awareness and attention. Famous tea master Yamagami Sōji, in his *Ten Rules for a Tea Master* (*Cha-no-yu-sha Kakugo Jittei*) writes:

> An encounter over tea is essentially *ichigo-ichie*, in other words, however many times the host and guest might meet over tea, this particular encounter today may never be repeated again, and so each encounter is a precious once-in-a-lifetime occasion.[4]

Ichigo-ichie means that life is all-too-brief, and each step on our journey, each moment, and each person we meet along the way is sacred. *Ichigo-ichie* not only embodies the spirit of Japanese hospitality, but also the essence of interpersonal, heart-with-heart presence. Akiko is living proof. As with Hiroko Suzuki's steadfast calmness and quiet strength, we find *ichigo-ichie*. Akiko and Hiroko exude their special presence and authenticity, what Zen Buddhists call *ishin-denshin*, a face-to-face, heart-with-heart connection. This goes beyond heart-*to*-heart. "To" indicates a false separateness.

Akiko is like the nightingale in Hans Christian Andersen's timeless story, "The Nightingale." She is real. Not like the artificial wind-up bird. And her story of beauty and truth, sadness and hope, are for me haunting lessons along our journey through ruins. My first nightingale, Hiroko, back in Tono, embodied lessons of steadiness and community. In contrast to Hiroko's quiet serenity, the diminutive Akiko exuded a vibrant, contagious energy. Small in stature, Akiko could move mountains. And she has.

Moments after the quake's angry and seemingly interminable growl split open portions of her tiny coastal village of Kamaishi, she spotted the receding sea. She remembered her grandmother's warning to quickly head to higher ground if this ever happens. Her inn was full. Akiko sprinted to each room and screamed for her guests to leave all their belongings and quickly run up the narrow path behind her ryokan. This was a path she had carved out years ago in anticipation of a tsunami. She cleared it regularly ever since. She bolted outside and shouted the same to her nearby neighbors, pleading for them to climb uphill. Those who heeded her pleas lived. Those who didn't perished, along with hundreds of villagers. Guests on the third floor who lacked the time and energy to run outside and uphill escaped to the top fourth floor instead. In a matter of minutes, the ordinarily majestic ocean, the tranquil scene all guests had from their rooms, transformed into a dark monster whose cold and wet jaw bit as high as the ceiling of the third floor. The monster licked the floor above while those on the fourth floor huddled together, some petrified, some praying. But they survived. Until the waters eventually receded days later, they lived on that fourth floor.

This was the same fourth floor where Kunio and I stayed during our visit in January 2012. As in all guestrooms and the dining hall, there is traditional *tatami* flooring. This is unique to Japan. Tatami is made from rice stalks dried into straw. The straw is then tightly bound together with thread and covered with rush on the surface. Tatami is supple yet firm, and effectively insulates against moisture. With my bad back, sleeping on tatami was my back-saver. Two years later, we volunteers slept on tatami mats.[5] With our room's traditional tatami flooring and *tokonoma*, a small alcove with hanging scrolls and flower vases according to the season, our room, on this top fourth floor, had a breathtaking view of the sea, now peacefully settled down after its furious eruption nine months earlier.

The fourth floor. Something about the number "four." The number four in Japanese, *shi*, is pronounced the same as the word for "death," *shi*. There is special significance to different words that have similar pronunciation, or homonyms. Roger Pulvers points this out when he compares the main character, Kenju, in Kenji Miyazawa's "Kenju Park Woods," with the author's first name, Kenji.[6] In each case, "Ken" is written differently but pronounced the same. Kenju and Kenji are alike in that they are both loners, less comfortable with human society, they prefer the company of nature and animals, and are often the subject of ridicule by others. Yet they are at home with nature. Kenji describes Kenju (his archetype):

> His eyes blinked and blinked in glee at the sight of the blue-green groves in the rain, he jumped for joy when he caught sight of a hawk flying up and up into the blue sky, and clapped his hands when he told everyone about it.[7]

As for numerical homonyms, *ku*, which means "nine," another inauspicious number as in 9.11, sounds the same as *ku* which also means agony and suffering. Combining "4" and "9," and we have an ill-omened number whose pronunciation can also mean "agonizing death." Japanese therefore tend to avoid using the number *shi*. Many hospitals avoid these as room and floor numbers. The fear of "4," what we clinically call "tetraphobia," is similar to the fear of "13," also shared by many Westerners, called "triskaidekaphobia." Numerologically, "13" adds up to the single digit "4." All Nippon Airways issue no seats with 4, 9, or 13. Under the skin of their hi-tech and consumerist society, Japanese have their fair share of phobias and superstitions. But fate is fate. Not a matter of numbers, at least on March 11, 2011. What matters is that all of Akiko's guests—those who scurried up the path and those on the fourth floor—survived due to her quick thinking, cool-headed anticipation, and diligence.

Akiko herself, after bounding outdoors and screaming to neighbors to head uphill, was soon swept up by waves. Her world turned upside-down, cold, wet, and black as she grabbed branches and twigs, submerged and struggling for air. She was choking, out of breath. She thought, "I'm going to die." Then, an unknown hand reached in, grabbed her, and pulled her out. Someone's hand. A man's. Some neighbor? To this day she does now know who saved her. He may have perished. Gasping to breathe, all she remembers is struggling to climb up to her fourth floor nearby.

The brutal waves, carrying their litter of overturned vessels, automobiles, and shattered buildings obliterated a major portion of the small town of Kamaishi. The sea claimed nearly nine hundred lives in Kamaishi, with over 500 still missing. Over 3,550 survivors lost their homes. Fatalities were higher in the port towns of Ishinomaki with over 3,000 deaths and Rikuzentaka's 1,500 victims. No doubt, because of Kamaishi's nearby surrounding hills, casualties would have been worse.

No Return to Normal

Akiko's ryokan reopened in January 2012, just before Kunio, Hiroko, Mizuki, and I arrived from Tono. Surviving neighbors and townspeople had been pressing Akiko to revive her inn earlier. But December was not the right time. For Japanese, December is a special month for families to be with each other. As with the United States and many other countries, December is a time for gift-giving. Shoppers deluge malls, and consumerism's tsunami targets major metropolitan centers like Tokyo, Osaka, and Nagoya. Life is relatively simpler in smaller cities and towns. And the coastal towns in the Tohoku region northeast of Tokyo, still reeling from 3.11, were experiencing a different December like no other. It was all the more important for family survivors to be with each other. Not only for each family, but for the community as a whole, as a collective family of survivors to commemorate the loss of members of their community family. Survivors who have lost their homes still live in temporary shelters that line up along major roads leading in and out of Kamaishi. December is the month to remind themselves of their strength in each other's company, whether families or strangers, in each other's in-betweenness. Indeed, there were and are no strangers.

Watsuji Tetsuro's notion of "in-betweenness" wears a deep metaphysical fabric. Recall his unique insights. In his pioneering *Fudo*, "Climate and Culture," he examines Japanese culture and other cultures in terms of their climate, and this climate embraces weather, landscape, and seasonal features, all combining to fashion distinct worldviews of nature, self, other, and their symbiotic entanglement.[8] Along with this meteorological feature, climate is also social. That is, we exist "in-between" other living beings. Our fundamental in-betweenness means that we are immersed in and shaped and colored by our natural and human geographies. Our essence is relational, not separate and private. *Aidagara* as in-betweenness does not denote a separation of disparate persons. There is no separate person A and separate person B, with intervening space as the in-between. Living in-between carries a more profound meaning, the in-between that *penetrates and permeates* both A and B so that space, in the form of meteorological and social climate, joins A and B together, defining each of them as ultimately interconnected. We are beings who relate to ourselves, others, and environment by virtue of how our surroundings relate to us. Space therefore, and this is crucial, assumes meaningfulness as *place*. *Aidagara* is about the power of place and our situatedness in place. For survivors, their temporary shelters become their relocated home, their shifted place. These same shelters will be in place for decades. At the same time, moving to these shelters does not guarantee healthy adjustment to a new place, nor does it ensure community. In-betweenness can surely lead to and sustain solidarity. Yet for those without family, spouse, children, land, home, and work, the sense of in-betweenness can also deepen that sense of estrangement that triggers despair, a despair that in turn can lead to suicide, as most likely the case with Akiko's daughter's best friend.

As for reopening Houraikan, for Akiko December was definitely not the right time. Families needed this month to be together. Houraikan officially reopened January 8, just before Japan's official "Coming of Age Day," celebrated nationwide on January 15. This is a time for young people to symbolically make the transition to adult status, with its new clothing of responsibility and duties. Young people legally become adults at the age of twenty, when they can now vote and marry without parental consent. I was in Tokyo to speak at the Global Alliance in Bioethics conference before traveling to Tohoku. There I witnessed hundreds of youth excited in preparing to celebrate their Coming of Age, with young men in their formal best and young women wearing traditional kimono.

For locals, the reopening of Akiko's inn was of great importance. It was a big step forward in lifting spirits and symbolizing hope after such tragedy. In the wake of 3.11, indeed after any major disaster, any so-called "return to normal" is a myth. Each hell-like unhinging shifts the social, economic, and political axis—all the more reason to reopen the inn. Houraikan's resurrection was indeed *the* news in Kamaishi. It meant far more than the much-needed kickstart for local businesses and tourism. Figure 4.1 shows us inside Houraikan just after its reopening and, from left to right, Kunio Aoki, Hiroko Suzuki, Akiko Iwasaki, me, and Mizuki Naito (now Abe).

For villagers throughout the area, their ryokan stands as a beacon of light and hope amidst death and despair. Akiko herself embodied this beacon of light in the weeks following that March 11 calamity. For months after the waters receded, they left a thick carpet of mud and debris. Throughout the region and coastline there was no power and no running water. There was no light. Just dark days and dark nights filled with a deafening silence. In the midst of this eternal night, Akiko discovered a working lantern with batteries in her ryokan. With her lit lantern, she religiously searched the area, stumbling through cracked streets, wading through mud and its dangerous wreckage, searching for neighbors in the hills. A few lone neighbors spotted her lantern from a distance. Light. An angel in the dark. One-by-one, she guided

Figure 4.1. Akiko (center) in Houraikan with, from left to right, Kunio, Hiroko, author, and Mizuki, January 2012

each of these survivors to the shelter of her still-standing but hollowed-out ryokan. For months, they nestled in her inn, making it through each day, together.

EARTHQUAKE-RESISTANT BUILDINGS?

An ancient legend has it that a giant catfish under Japan moves in ways that creates earthquakes. In doing so, the catfish doles out good and ill fortune to redistribute the wealth that comes from class division.[9] The catfish is an equalizer. So too is Nature, life-giving as well as destructive, our original equal opportunity employer. In the end, Nature has the last say. She has a mind of her own. Throughout their history, Japanese have learned that the hard way. We cannot rule Nature. We can only *live along with Nature*, befriend it.

So, why was Akiko's inn still standing? It was surely a stroke of providence. Akiko's forethought factors in as well. While all other surrounding buildings were crushed, collapsed, or swept out to sea, her ryokan remained upright. The Great Hanshin Earthquake struck the major city of Kobe on January 17, 1995. The quake weighed in at magnitude 7.2 on the Richter Scale, caused widespread fires, and took nearly 6,500 lives. Akiko will never forget listening to victims' screams on her radio. Since then, she vowed in her heart to do whatever she could to protect her guests from future jolts. For months, she and her staff cleared ground, cut down trees, and created a path on the hill behind her inn. This was exhausting enough. To add to this, she nearly exhausted her funds and savings to ensure for her guests a more earthquake-resistant dwelling.

With Japan's island archipelago resting on top of the earth's major fault lines, this need to construct earthquake-resistant buildings has a long history. Japanese carpenters, *daiku*, were also trained as both architects and engineers. They designed their traditional wooden buildings with earthquakes in mind. Heavily mountainous, Japan has an abundance of wood for use in structures, and, given certain wood's flexibility, pliancy, and strength, *daiku* designed buildings in ways that, blended with proper engineering, would better withstand an earthquake's rocking motion, what we call "seismic," or "earthshaking." The more "scientific" study of earthquakes surfaced after the Meiji Restoration, beginning in 1867 with the government's surge of interest in things Western. Here, we enter a fascinating chapter in Japanese architecture which blends technology, engineering, science, and art. Recall that, as in the case with the word for death, *shi*, and its similar sounding *shi*, meaning "four," words with similar pronunciations, homonyms, take on special meaning for Japanese. Take the word for "technology," *gijutsu*. This sounds more like *bijutsu*, the word for "art," than it does *haguku*, "science."[10] This is quite revealing in a technologically saturated society like Japan. Furthermore, foreign scientists who spurred on the full-fledged science of seismology that developed soon after the Meiji Restoration were from countries *not* accustomed to earthquakes. These scientists, chiefly British, were primarily responsible for establishing Tokyo's Seismological Society of Japan (*Nihon Jishin Gakkai*) in 1880. Prior to the Meiji Restoration, there was little obsession with earthquake study. Being unhinged was a part of the natural order. Now, the science of seismology displaced the catfish. With increasing awareness of the country's seismic activity, "experts" became more aware of this place as a land of constant tremor.

Let's travel back over a century to the devastating Great Nōbi Earthquake on October 28, 1891. Nōbi, the region north of Nagoya covering a wide swath of Honshu from Osaka to Tokyo experienced its terrible death rattle, one felt throughout the country. This was in 1891, well before the invention of the Richter scale. Experts, however, place the magnitude comparable to 8.4, the strongest measured quake in Japanese history until March 11, 2011. In 1923, the Great Kantō Earthquake that demolished Tokyo and Yokohama measured 7.9. The Nōbi quake took over 7,300 lives, injured tens of thousands, and left more than 100,000 homeless.

The Meiji Restoration prompted the learning of matters Western in law, government, economy, and architecture, particularly the science of building earthquake-resistant structures. Brick, stone, iron, and steel began to replace traditional wooden structures and bridges. In his superb investigation of the relation between Japanese architecture and earthquakes, *Earthquake Nation*, historian Gregory Clancey reminds us that this "technology transfer," a result of having Japan's doors prodded open by Western powers when Commodore Matthew Perry's blackships appeared in Tokyo Bay, was less a transfer of techniques but more one of knowledge and imbedded beliefs. One such imbedded belief was the Western idea that brick buildings were *per se* stronger than wood and could thereby better resist the force of powerful quakes. Newfound masonry and iron represented strength, whereas traditional temples, pagodas, and homes made of wood and bound together with ropes appeared more delicate, fragile, and worse, antiquated.

The Great Nōbi Earthquake proved this to be wrong. In an earthquake, the risks of death come not so much from the ground swallowing up helpless victims, as in Hollywood versions, death from the bottom up. Deaths come mainly from collapsed structures raining down on victims from above—death from the top down. Most of Nagoya's "stronger" constructions collapsed, many of them government buildings, railways, steel bridges, and brick buildings. Pagodas, temples, and many wooden houses, on the other hand, withstood the powerful, lingering shocks. For instance, Nagoya's new stone post office collapsed while the nearby wooden ryokan survived.[11] Nagoya's famed castle also stood upright in contrast to surrounding crumbled brick buildings and fallen telegraph wires. By the way, despite its having weathered earthquakes, Nagoya Castle burned to the ground during the American military firestorm in World War II. Nōbi's devastation sparked further critique of Western technologies like new water pumps and oil lamps, much of which exacerbated the ruin.[12] It also underscored the link between building construction and catastrophe.

Actually, this was becoming apparent prior to the Nōbi earthquake. In February 1880, a large earthquake in Tokyo toppled numerous tombstones of stone and masonry, and cracked buildings throughout Tokyo's Ginza section, a hugely popular upscale shopping district, much of it influenced by European architecture. For English geologist John Milne, these same buildings, due to direction and depth of their cracks, became ready-made seismometers.[13] We can learn from toppled gravestones. And consider the horrible 1908 calamity in Italy. On December 28, Europe suffered its worst earthquake. In less than one minute, in Calabria and Northern Sicily, nearly 120,000 people were crushed to death, Messina and Reggio were flattened, and 98 percent of all structures collapsed. Almost all buildings in the region were masonry. In comparing the fatalities in Messina and Nagoya, Japanese seismologist Ōmori Fusakichi pleaded for proper construction that would ride a seismic wave, building a case for wood.[14]

Earthquake's Deadly Sibling

Are we over-rating wood? The burning of Nagoya Castle cautions us about over-generalizing. Another lethal feature in earthquakes, one against which traditional wood buildings stand little chance, is fire. Actually, most fatalities usually come from ensuing fires spreading all-too-quickly in wooden homes. Though wooden structures may be pliable enough to sustain the ground's angry motion, they do not survive the raging conflagration, an earthquake's ugly fraternal twin.

Wood undoubtedly assumes natural significance in Japan, often described as a "wood culture." Japan is not alone. The same can be said for Norway with its immense proliferation of forests. My Norwegian friend and colleague Professor Julie Feilberg, who teaches applied linguistics at Norwegian University of Science and Technology in the wonderfully quiet town of Trondheim, tells me of the highly complimentary term *hel ved* to describe a person who is genuinely forthright and sincere, a person of good character and integrity. *Hel ved* literally means "true wood," wood that is strong and enduring. When I met with Professor Are Holen, who, now retired, taught at Trondheim and Oslo, I knew what she meant. Are is a renowned psychiatrist. He is also an intellectual icon as a pioneer in introducing Hindu meditation after his long experience and study in India. Are is also well-known for his long-term study of the effects of trauma on survivors of the March 30, 1980, collapse of the Alexander Kielland oil platform by strong waves off the coast of Norway. Of the 208 oil workers the platform had housed, 123 died.

Are was a leading expert witness who testified during the long, drawn-out trial of Anders Behring Brevik, the lone mass-murderer of over seventy young people at that ill-fated summer camp just outside of Oslo, July 22, 2011, just over four months after Japan's calamity. Are, to me, embodies *hel ved* in his standing by his beliefs and principles throughout the trial, arguing that the mass murderer was indeed suffering from a deep-rooted psychiatric disorder, psychotic paranoid schizophrenia. The defendant and others experts believe he was perfectly sane and of sound mind.[15] He was found guilty of premeditated murder and remains in prison.

As to this compelling relationship between wooden structures and seismic activity, consider the durable resilience of the pagoda. Japan's oldest pagoda is the five-story edifice at Hōryūji, built around the late seventh century. The British engineer Christopher Dresser was astounded to discover that the pagoda's center of gravity was sustained during a quake due to the fact that its huge central wooden mast acts as a pendulum.[16] American architect Ralph Adams Crane concurred with Dresser, claiming that Japanese architecture represents the "most perfect wooden style the world has known. As such it must be judged, and not from the narrow canons of the West, that presuppose masonry as the only building material."[17]

Although the Nōbi earthquake planted a deep vein of doubt and suspicion regarding opening doors to a more "Western landscaping,"[18] Japanese architects and engineers also realized the need for a more practical, efficient, and safe compromise between East and West. The quake's aftermath left cities, towns, and villages in heaps of wood and stone, a completely crushed body with bone bits both broken and pounded into pulp. Widespread fatalities came not only from heavy, falling objects, bricks, and steel but especially from infernos that consumed wooden structures and their inhabitants within seconds. This was the inferno retired truck driver Bunzo Kanayama witnessed on March 11, 2011, blazing throughout his town of Otsuchi. At the same time, he saw massive waves sweep away his home and his neighbors. We will meet Bunzo later on. As to wood, there is a trade-off. Wood unhinged by a shaking earth may not cause the kind of injuries from falling concrete, bricks, and steel. Yet seismic waves and powerful, towering sea waves ignite raging fires which swiftly devour wood.

Not all scholars embrace the idea that wood had been used deliberately to palliate the effects of seismic activity. Sano Toshikata tells us that wood was used primarily not only on account of its abundance, but also because it represents simplicity and naturalness in harmony with Japanese tastes.[19] In other words, wood embodies deep cultural meaning. For Akiko and other survivors in Kamaishi, however, the thin pine-tree-filled wooded hillsides provided literal shelter and escape from their monster.

Did building construction make a difference in later earthquakes? The Great Kantō Earthquake, September 1, 1923, shook Tokyo, Yokohama, and surrounding areas to its depths and spawned unparalleled horror, shattering over 10,000 buildings, and taking around 140,000 lives, most of whom perished in flames that devoured both abundant wooden and modern structures. Since then, brick buildings were banned in favor of ferro-concrete structures, combining steel frames with concrete. Compromise came in the form of frames with a wooden core layered over with a masonry face, similar to earlier work laying earthen plaster over wood during the Edo period. This enabled better fire-proofing.[20] Like the Tohoku disaster, the earthquake itself triggered further natural and technological destruction. Tohoku's population density, however, did not compare to that of Tokyo, and this played a major part in the Kantō destruction. A simple formula—the more people, the more deaths. Nonetheless, Tohoku's triple horror of earthquake, tsunami, and nuclear meltdown made it the nation's worst calamity since Hiroshima and Nagasaki.

HIGHER SEAWALLS?

Despite its fragile topography resting on top of the earth's deep-rooted fault lines, despite its enduring history of calamity and countless casualties, Japan has managed to both survive and to move onward. Our Western knowledge, *yōgaku*, as to taming nature is not enough. To this day, we cannot with certainty predict the when, where, or magnitude of earthquakes and tsunamis. A disturbing paradox becomes increasingly evident. We Americans continue to expend high-priced efforts to tame Nature. But as our ubiquitous screen-life has taken over, we've become increasingly indifferent to Nature. The screen has become our substitute for what is real. For many of us, sheltered from the storms that thrash those "others over there," we live increasingly in our screens. The screen has become the new real. It outreals the real. We live in the world of the "hyperreal." As philosopher Jean Baudrillard (1929–2007) uncompromisingly puts it in his *Simulations*:

> The very definition of the real has become: that of which it is possible to give an equivalent reproduction . . . The real is not only what can be reproduced, but *that which is always already reproduced*: that is the hyperreal.[21]

The 3.11 earthquake and tsunami literally obliterated towns and villages along the Sanriku coastline. People there now face rock-hard choices. Should they build higher seawalls? In the popular resort town of Rikuzentakata, where I volunteered to work in the shelters, the seawall height of twenty-three feet did little to block the initial tsunami height of 58 feet, with some waves as high as ninety feet. Out of its population of 24,246, Rikuzentakata lost 1,732 of its citizens. And at Minamisoma in Fukushima Prefecture, where I met with Mayor Sakurai, its twenty-foot high seawall was not enough for the waves that rushed in at nearly forty feet.[22] Should people move farther back from the seacoast? As to building higher levees, concerns in Kamaishi are similar to other towns. As a member of the village council, Akiko, along with many others, remains staunchly opposed to constructing higher concrete levees to protect the village against future waves. The small, ever-smiling, and energy-packed ryokan owner shared with me her reasons, all of which are echoed by other survivors. Akiko tells me "We choose to live here, to live by the sea." Villagers do so because they love the sea, its beauty, peacefulness, and sense of connection. They know full well the risks involved like typhoons and tsunamis. They willingly take these risks. Moreover, for many villagers, higher seawalls are both ecologically damaging and aesthetically unattractive. Materials used for those seawalls—concrete—take their toll on the environment, both human and maritime. And diminishing people's views of the stunning seascape, they detract from the beauty of living by the sea.

Most importantly, we humans can never control nature. On March 11, massive waves swept over the stoned, walled defenses protecting the Fukushima Daiichi Power Station and spawned a lethal nuclear meltdown and spread of an invisible enemy—radiation. Colossal waves overcame Miyako City's "Great Wall" sea wall barrier built to protect against tsunamis. A formidable earthquake can just as easily rupture any levee. For Akiko and her townspeople, instead of attempting to control Nature, it is a matter of living with Nature and educating ourselves and our future generations about what to do in a crisis, in this case, to run to higher ground. Akiko attests to this when she implored her guests to either move to the top floor of her ryokan or climb up the hill behind her inn. The abundantly wooded pine-tree hills surrounding Kamaishi is Nature's gift, supplying villagers with a means to deal with her dark side.

Akiko and her townspeople are not alone in their opposition to building higher seawalls. Historian Norio Akasaka, a native of Tohoku, still lives there. He is an expert in Tohoku culture and history. He also desires to keep in sync with nature. Akasaka reminds us that many of the coastal towns were built on what were originally natural lagoons. Having towns built on these allowed fertile area for rice-faming, which requires wet ground. Akasaka believes that towns need to retreat farther back and restore that land to its natural setting.[23] This means deferring the need to build higher and more costly seawalls. Kamaishi's extensive seawall, which was easily breached on 3.11, was exorbitantly expensive, costing around $1.5 billion.[24]

Shigeatsu Hatakeyama was one of the United Nations' "Forest Heroes" award recipients. "Forest Heroes" are those who demonstrate a genuine commitment to restoring forests. He too lives in Tohoku and also argues against building higher seawalls. His reasons are again environmental—it will be harsh and artificial severance of the natural link between sea and mountains. Along the U.S. northwest coastline, this link between sea and mountain constitutes its stunning natural beauty. Moreover, the construction of higher seawalls would adversely affect marine life. He states, "To us, it seems far more important to think about how to get along with the sea and the mountains than to consider how to deal with the threat of tsunami."[25]

Daniel Aldrich and Yasuyuki Sawada convincingly show that these seawalls offered less impact on mortality rates than what they consider degrees of internal social bonding.[26] A community with a relatively high degree of social cohesion in that neighbors habitually look out for each other, particularly the more vulnerable—elderly, those who live alone, and families with small children—is more likely to alert each other and these vulnerable populations of the danger of waves sweeping over their "safe" walls, and would more likely strongly urge and help each other to run quickly, together, toward higher ground. This was clearly the case with Akiko who took it upon herself to commandeer and organize her guests to either ascend to the top floor of her inn or run immediately up the hill behind her ryokan. Through her actions, all of her guests were saved. Along with Akiko and many others in her town, Aldrich and Sawada recommend that, rather than expend exorbitant sums on physical infrastructure like seawalls, more funds should be allotted to creating incentives for building and enhancing community, to strengthen community ties, and engaging people of all ages. They argue that this should be the thrust of public policy measures regarding disaster. Akiko does not stop there. She has plans for the future. She knows the value of community. She knows how essential it is to keep the community together. Her dream is to sustain the tight-knit bonds of her local community by using these hills as geographical, physical, and culturally appealing "escape" routes. She envisions widening the path behind her guest house and establishing similar paths by other volatile areas throughout the town. They would have rest stops along the way. Some would even have small ampitheaters for picnics, music and theater performances. This would enable people to live closer to nature in safer zones. But all this requires both education and official and fiscal support.

Many who survived the tsunami did so because they knew how to respond—head to higher ground. Many who died did not respond accordingly, and instead headed to official evacuation sites for earthquakes. They died in these sites. For Akiko, the key lies in education, learning more about earthquakes and tsunami physics and how to respond. An inspiring example are the children at East Kamaishi Junior High School.

Students Save Lives

Throughout the country, Japanese children learn about earthquakes throughout their schooling. This education begins at kindergarten and continues through high school. Students learn about the physics of earthquakes, their causes and effects, evacuation guidelines and safety, interpreting earthquake warnings, respectful evacuation behavior, and proper response to minimize injury. Their learning is put into practice on the first day of each new school year with a required earthquake drill. Not only are students more informed, but they are also trained to hopefully exercise the right mental attitude should an earthquake occur. So also with tsunami education. Knowing what to do in a tsunami demands learning from the past, such as lessons from the terrible 1896 and 1933 tsunamis. Kamaishi was one of the towns taking a more strident approach to being properly prepared, which remains the key to disaster survival. This also means heeding the studies of experts who predict the next Big One soon. The town council invited Gunma University's disaster expert Toshikata Katada to help train elementary and junior high school students about tsunami evacuation. After his visit in 2005, schools then complemented students' training with education about tsunami physics and history in their classes.[27]

In Kamaishi, 3.11 took 1,046 lives. Out of the town's 2,900 elementary and junior high school students, five perished.[28] Why only five? Here's what happened. In Kamaishi, students, now having been taught the physics and history of tsunamis, regularly practiced evacuation drills. The magnitude 9 earthquake knocked out power and electricity, disabling tsunami public warning systems. This did not stop students from knowing what to do. Soon after evacuating to their third floor as instructed, young students and teachers at Unosumai Elementary School spotted students at nearby Kamaishi East Junior High School outside, running towards a designated evacuation site. The site was on slightly elevated ground. Some teachers at the elementary school insisted that students who were watching stay indoors in their classrooms. Other teachers at the junior high school instructed their students to stay inside as well. Yet despite their teachers' instructions, students at the junior high school instead ran to the nearby shelter. Immediately after the younger children watched them head outside to the evacuation site, they intuitively ran after the older students and left the elementary school. When they all reached the site, the older students realized that the elevation was not high enough. They decided to run up towards higher ground further away from the site. Their lessons from tsunami education and training stayed with them. They ran to high ground. More importantly, these students did not run helter-skelter, each for him or herself. *They stayed together, guided by the older students.* They were all in this together. By that time, there were elementary school children, high schoolers, teachers, and elderly. In scrambling their way up the nearest hill, all junior high students grabbed the hands of younger children and elderly, carrying many of them on their backs. All students and teachers from the two schools survived—350 elementary school children, 212 junior high schoolers, and twenty-one teachers.

With Nature as teacher, Japanese seem more prone to embrace life's fragile transience. With its heritage of earthquakes, tsunamis, fires, and typhoons—nature's mixture of earth, air, fire, and water—this leads to an acceptance of what is natural, an acceptance that some would call "fatalist," even boastful of inviting disaster. Fate in the form of chance takes on a particularly curious meaning in a culture devoted more to embracing our temporariness rather than clinging to some illusion of permanence. Many Japanese seem to have a particular affinity towards matters of fate or chance, like eating the potentially poisonous blowfish, *fugo*, served as an expensive delicacy in top restaurants. Some well-known figures and celebrities who have tasted *fugo* have died from their roll of the dice. Akiko and her community's opposition to levees, however, does not reflect fatalism. Rather, it illustrates what many Japanese view as a vital prerequisite on this brief journey through life, a need to *pack lightly*. Only if we pack lightly can we nurture change, not permanence. This is the long-cherished appeal of the cherry blossom, *sakura*, whose beauty, like ours, lies precisely in the abiding truth of its short life and fleeting, precious presence.

NOTES

1. Lucy Birmingham and David McNeil, *Strong in the Rain: Surviving Japan's Earthquake, Tsunami, and Fukushima Nuclear Disaster* (New York: Palgrave Macmillan, 2012), 54.
2. Michael C. Brannigan, "Our Nightingales' Song from Tohoku," *Japan Railway & Transport Review* 60 (October 2012): 60.
3. Akiko has become somewhat well-known since the calamity. In the aftermath of 3/11, Akiko has become a spokesperson for the resurgence of her community by attending international meetings and sharing her story of renewal.
4. At http://www.ryokan.or.jp/english/pdf/glossary_of_terms.pdf; accessed March 5, 2013.
5. The tatami mat is often used as a measuring unit. The floor space of a room is measured according to its number of mats.
6. Kenji Miyazawa, *Strong in the Rain: Selected Poems*, trans. Roger Pulvers (Northumberland, UK: Bloodaxe Books, 2007), 20.
7. Kenji Miyazawa, "Kenju Park Woods," cited in Kenji Miyazawa, *Strong in the Rain*, 20.
8. *Fudo* was originally published in 1935, taken from notes Watsuji had written in the 1920s. It was translated into English in 1961. There's a curious omission in Watsuji's study of the relationship between climate, geography, environment and national character. He makes no mention of earthquakes. He discusses the meteorological phenomena of rain, humidity, and typhoons, emphasizing air, water, temperature, and stressing the centrality of humidity in Japan, but not its landscape. In his earlier work, *Pilgrimages to the Ancient Cathedrals in Italy*, he even compares Japan and Southern Italy, but again without mention of earthquakes. See Gregory Clancey, *Earthquake Nation: The Cultural Politics of Japanese Seismicity, 1868–1930* (Berkeley: University of California Press, 2006), 232.

9. Woodblock prints of this giant catfish, *namazu-e,* were quite popular after the Great Ansei Earthquake in 1855 that shattered nearly all of Edo, the ancient capital of Japan which later became Tokyo.

10. Clancey, *Earthquake Nation*, 2.

11. Clancey, *Earthquake Nation*, 118, note 272.

12. This is similar to earlier opposition of introducing the gun to Japan, a weapon considered by many Japanese to be a cowardly threat to the cultural and historic symbolic purity of the sword. The use of the gun in Japan was short-lived. See Noel Perrin's excellent account in his *Giving Up the Gun: Japan's Reversion to the Sword, 1543–1879.*

13. Clancey, *Earthquake Nation*, 73–74.

14. Clancey, *Earthquake Nation*, 172.

15. Interview with Are Holen, Oslo, June 20, 2012.

16. Clancey, *Earthquake Nation*, 51.

17. Cited in Clancey, *Earthquake Nation*, 52.

18. Clancey, *Earthquake Nation*, 62.

19. Clancey, *Earthquake Nation*, 232.

20. Along these lines, in an interesting case, the American Fuller Company designed the Maru-Biru, or Marunouchi Building using steel frame and insisting on building it "the American way," rather than relying on the Japanese ferro-concrete. Tokyo's smaller earthquake on April 26, 1922 damaged the still-under-construction Maru-Biru. After this, the builders blended concrete with the steel frame. Despite this, however, the Maru-Biru, as with many other modern buildings, suffered extensive damage. See Clancey, *Earthquake Nation,* 220.

21. Jean Baudrillard, *Simulations* (New York: Semiotext[e]), 1983). 146.

22. Birmingham and McNeil, *Strong in the Rain*, 50–51.

23. Birmingham and McNeil, *Strong in the Rain*, 182.

24. Norimitsu Onishi, "Seawalls offered little protection against tsunami's crushing waves," *New York Times*, March 13, 2011, accessed January 27, 2015, http://www.nytimes.com/2011/03/14/world/asia/14seawalls.html?pagewanted=all&_r=0.

25. Hideyuki Miura, "Healthy Forests Key to Rebuilding Tohoku Communities," *Asahi Shimbun*, April 3, 2012, cited in Birmingham and McNeil, *Strong in the Rain*, 183.

26. Daniel Aldrich and Yasuyuki Sawada, "The physical and social determinants of morality in the 3.11 tsunami," *Social Science and Medicine* 124 (January 2015): 66–75.

27. Birmingham and McNeil, *Strong in the Rain*, 38.

28. Setsuko Kamiya, "Students Credit Survival to Disaster-Preparedness Drills," *Japan Times*, June 4, 2011, accessed January 27, 2015, http://www.japantimes.co.jp/news/2011/06/04/national/students-credit-survival-to-disaster-preparedness-drills/#.VMeuQmctDcs. Also cited in Birmingham and McNeil, *Strong in the Rain*, 39.

Chapter Five

Community and Connectedness: Bunzo and Jin

What is deadlier? Baptism by water or fire? Immediately after the tsunami, fires swamped numerous coastal towns. From a distance, Bunzo Kanayama, a retired truck driver in his sixties who lives in Otsuchi, watched helplessly as waves swallowed his home and neighborhood. He also watched as fires raged throughout his town. He spotted a powerful wave crushing storage tanks holding petroleum. The wave shattered gasoline tanks, smashed cars, and spilled tons of fuel. Vessels of all sizes were swept ashore, colliding with overturned vehicles, and their oil and gasoline emptied into the sea of mud. Once the gasoline and oil burst into flames, the onrushing sea's hell turned into a raging inferno. Nearly everywhere, gasoline fires exploded. Hundreds of villagers rushed to one of the designated shelters, a popular Buddhist temple about a half mile inland from the coast. They no doubt trusted in the Buddha's protection. That same temple saved them in the past. Yet no one was prepared for a tsunami of this massive force and height. They probably didn't know what hit them as the giant wave easily crushed them to death. Within seconds their temple became a fiery hell. Numerous casualties throughout the coast came from those who sought shelter in these designated evacuation sites.

FETCH ME THE RAINLIKE SNOW

Bunzo's *Nanohana*

In the months following the loss of his home, relatives, and friends, Bunzo would not allow himself to be paralyzed by grief. He felt a burning desire to do something that would inspire survivors. He was determined to plant *nano-*

hana, or rapeseed flowers, throughout the ravaged areas. Why *nanohana*? No doubt, the most popular flower in Japan is the *sakura*, the pink cherry blossom and one of Japan's symbols. *Sakura* bloom in Spring and last only a few days at their peak, reminding us of the fleeting nature of life. *Sakura* symbolizes the precious beauty of our brief lives. Yet prior to the full bloom of *sakura*, stunning yellow carpets of *nanohana* fill areas by roadsides, riverbanks, and open fields.

Nanohana literally means "vegetable flower." We call the *nanohana* plant canola, from which we extract canola oil. *Nanohana* is a healthy vegetable closely related to broccoli, and Japanese have long used *nanohana* as food, vegetable oil, and animal feed. During the Edo Period (1600–1868), the plant's oil was used to light lanterns, thus later symbolizing light amidst the coming of dusk and darkness. Yearly, thousands of tourists flock to Japan's largest seeded area of *nanohana* to witness *Nanohana Matsuri* (Rapeseed Blossom Festival) in Yohohama (where my sisters, identical twins, were born). In Tokyo, many also visit the Hamarikyu Garden in Shimbashi with its open field of *nanohana* blossoms.[1] They remain a striking harbinger of Spring after a cold and bleak Winter.

Their yellow-color is especially significant. Rapeseed flowers are known for their brilliant yellow, and, particularly in Buddhist ceremonies, the color yellow symbolizes hope, or *kibō*, and life. Of the five Buddhist colors—blue, black, red, green, white, and yellow—yellow has the highest symbolic value. Yellow comes closest to representing daylight. Moreover, its color is close to the saffron robes worn by monks and thereby represents renunciation, humility, and detachment from materialism. As the color closest to earth, on the one hand it symbolizes rootedness and restraint, reminding us to control our desires. On the other hand, it stands for sustenance, reminding us of nature's nourishment. In the aftermath of 3.11, yellow represents new life and hope.

For Bunzo, *nanohana* would be a sign of renewal to survivors. They would remind survivors that life still goes on, that the future holds new beginnings. As Hermann Hesse writes in his "Stages," "In all beginnings dwells a magic force, for guarding us and helping us to live."[2]

At first, Bunzo was a solitary figure daily planting his rapeseed flowers, despite wind, snow, rain, and cold. Townspeople wondered why this old man was out there, putting up with the bitter cold and wet. Had he gone mad? Soon, a bare handful of young people joined him. The handful grew larger as they spread the word—"Join us in helping this old man plant flowers." In weeks, a small army of survivors of all ages, mostly young, toiled together with him in planting his seeds of hope.

When I first met Bunzo, his unforgettable smile seemed permanently etched into his ruddy face. Now, Bunzo's dedicated company of volunteers continues to grow. People from all over Japan come to work with him in his rapeseed fields. He now organizes an annual rapeseed flower festival in late

Spring to honor the victims of the disaster and to celebrate the promise of renewal and hope. (Just before I left for Oslo, Hiroko sent me a photo she had taken during her visit to the festival.) And, most importantly, he celebrates this renewal and hope through Nature. It was Nature that crushed his town and people. And it is through being one with Nature rather than fighting Nature, laboring together planting seeds, that Bunzo and his committed followers find redemption. His rapeseed festival reminds us of how we find hope through Nature.

A volunteer with whom I worked, Yumiko Murakami, was quite familiar with Miyazawa Kenji's life and works. Hiroko had already introduced me to Miyazawa, and I was eager to learn more. In her forties, Yumiko has volunteered every year since 3.11. Wonderful and gracious in so many ways, Yumiko and I would often talk about Miyazawa's plays, shorts stories, and poems. One day, she took time off from work, traveled by train to Hanamaki, Miyazawa's birthplace, and returned that evening to share with me some of his poems and other writings.

Miyazawa Kenji (1896–1933) was the oldest of five children. His father, Masajiro, was Hanamaki's pawnbroker, a necessary profession in those days of constant hardship in Iwate Prefecture and throughout Tohoku as people struggled each day to get by. Kenji broke tradition by not taking over his father's business. Instead, his studies turned him into a devoted follower of Nichiren Buddhism. His father belonged to the opposite, quieter Jodo Shinshu Buddhist sect. Kenji more closely studied agronomy and had a variety of interests, ranging from collecting rocks, learning astronomy, to playing classical music on his cello. He eventually left teaching agronomy to become a farmer.

Of all Japan's modern poets, Kenji Miyazawa most untiringly reminds us of our natural affinity with nature. For him, Nature *is* Life. He persists in reminding us that we humans are not the measure of all things. Nature is the measure of all, in its fascinating, miraculous, beautiful, and at times terrifying unfolding. March 11, 2011, illustrated this horrifying power. It is no surprise that the interest in Kenji Miyazawa is so timely. His ideas represent a refreshing departure from Japan's mounting tidal wave of modernization, industrialization, and materialism ever since their sparks in the Meiji Restoration (1868–1912). This gradual storm surge of consumerism brought about a neglect of older, traditional values regarding our all-important kinship with nature and the natural world. For Kenji, ignoring our natural kinship slices away our humanity, our spirit. Hence, throughout his writings, incessant voices of nature—trees, plants, animals, birds, insects, wind, sky, sea, and earth—whisper to us, reminding us they are ever-present.

For Kenji, this is not merely a matter of abiding by nature in passive resignation or simply giving oneself over to her. For him, this means actively embracing nature, to engage her in a self-offering. We see this self-offering in his dying sister's last words. His sister Toshi died of tuberculosis at twenty-four. Being there with her in her last moments left the most profound impressions upon his heart. This was no doubt his most traumatic personal earthquake. In his beloved poem, "The Morning of Last Farewell," he describes Toshi's final moments using the imagery of that day's incessant sleet and her final requests for him to "fetch me the rainlike snow."[3] Nature's voice stubbornly reminds us of our finitude. And Kenji habitually describes the weather outside to reflect its kinship with the personal weather inside. Yet, for Kenji, death has no finality, and life is a mere spark on our journey. His poems repeat his never-ending message—the secret of life lies in living peacefully with nature. If we do not, we die in spirit. This is the secret I always relearn when visiting the sea, which is probably why I love the sea so much and view it as my lifelong teacher. What the sea constantly reminds me is that no matter how much anguish we encounter in our small seas of anguish, "Compared to the great virtue that envelops the world" all of our suffering is, as Miyazawa puts it, merely "a little drop of dew on the point of a thistle's thorn."[4]

Bunzo's rapeseed festival reminds us of that secret. For the people of Tohoku, Miyazawa has become their literary voice of hope after the disaster. His works have inspired the people of Tohoku to brave the tragedy and move on with promise for the future. In the same way, Bunzo's flowers plants seeds of hope in survivors' hearts.

Along these lines, but with a different intention, there is a farmer in Iitate Village in Fukushima Prefecture, just over thirty kilometers (18.6 miles) from the Daiichi nuclear plant, who decided to remain in Iitate even though it is now mostly abandoned after villagers were told to leave due to radiation. Nobuyoshi Ito questions the legitimacy of official radiation levels in the area, arguing that it is not as unsafe as authorities claim. He is there to prove his point, even though he is officially registered at a shelter in Fukushima City. He only goes to the shelter to collect mail and spends his days caring for his small farm of vegetables, eggplants, rice, and potatoes, not for consumption but to test radiation safety levels.[5] Furthermore, Nobuyoshi intends to plant thousands of acres of sunflowers. Nobuyoshi hopes that his sunflowers may help soak in radiation, though he has little confidence that they do. As to claims that sunflowers may soak up radioactive cesium, whether or not sunflowers can effectively absorb radiation is arguable. More importantly, however, for Ito, like Bunzo, he hopes that his flowers will offer beauty and the promise of rebirth in what is now a barren landscape.[6] Nobuyoshi's message is similar to Bunzo's. The victims' spirits live on in survivors' hearts and in Nature, and the region will live on in stronger bonds of community that cannot be broken.

Beyond any doubt, Bunzo's rapeseed project inspires hope, *kibō*, among many survivors. This link between hope and having a project, aiming towards a future of growth, is underscored by Genda Yūji, founder of "hope studies," *kibōgaku*, at Tokyo University. He traveled with Uno Shigeki to Kamaishi.[7] Genda brought with him calendars. Calendars represent the inseparable relation between time and hope. The September 2011 protests in Tokyo echo this calendar motif with the ongoing slogan, "Save our children," *Kodomo mamorō*, that is, "Save our future, for hope resides in our children." Children embody the future. There needs to be a future project, like planting flowers for their blossoming. One must have a future project in order to situate oneself a step ahead. Envisioning oneself in the future becomes the ground for acting and living fully in the present. In the aftermath of calamity and crisis, the future project becomes the ground for the present, not the reverse. I well remember our dad's small calendar hung over his tiny kitchen table was filled with events, arrivals, departures, doctors' appointments, birthdays, anniversaries, and so forth. Like us, calendars are inherently future-oriented, and for Dad, they kept him moving and alive each day, especially after Mom's passing. Hence, a bold awareness of *aida* in time, of our natural situatedness towards the future, so profoundly described by Heidegger, can help pull us out of present misery with a future project. My lifelong colleague and friend Joe Cunneen, whose cancer was eating away at him, always reminded me of this in his last months, "I need a project." Having a project helps redeem our present anguish.

One evening in March 2014 when I was volunteering at Tono Magokoro Net (more on this next chapter), my friend Hiroko, another volunteer, Oka, a student with a chronic stutter, and myself went to Tono City for dinner. At a local restaurant, we were joined by Katsunori Kawano and his colleague Naoki Hayashi, who both work for Fuji Xerox where Katsunori is a manager in its Communication Technology Laboratory. With the music of Abba curiously playing in the background ("Dancing Queen"), Katsunori and I discovered a shared interest in the Japanese Zen Buddhist Dogen (1200—1253), perhaps Japan's most influential philosopher and who founded Japan's Soto Zen school. Watsuji Tetsuro was profoundly influenced by Dogen's teachings and wrote a leading commentary on his ideas. Katsunori also shared with us his dream project of establishing a Center for Hope (*Kibō*) in Tono. This would be a center primarily for youth, with students from elementary schools and high schools. These youth are Japan's future, and Katsunori intends to have his center for interdisciplinary studies instill and cultivate hope for the future among the youth amidst the ashes of 3.11. He and I stay in contact, and his center is now underway.

Fishing's Slow Demise

When I revisited Bunzo in winter 2014, he, Hiroko, myself, and two of his friends ate lunch at a nearby *kasetsu* seafood eatery. Up until March 11, Otsuchi, like many coastal towns, has had a long history of fishing. And, like most fishing towns, it prided itself on its catch. The town was known especially for its abundant catch of flounder, shellfish, abalone, sea squirt (scallops, called *hoya*), and popular seaweed like *nori* and *wakame*. The owner of the *kasetsu* eatery, a fisherman, boasted of his catch of sea squirt, which I ordered. It was the biggest sea scallop I've ever eaten. The owner proudly showed us his collection of giant scallop shells, embossed in their rough, dimpled designs. The owner looked to be in his late fifties, young compared to most of the fishermen who fish along the Sanriku coast. A significant number of fishermen are easily in their sixties. This aging cohort faces a giant economic hurdle in the future as more fishermen will die off, unless young family members are willing to take on what may become a dying profession.

This maritime region once thrived on commercial fishing. The Sanriku fishing industry's impact was global. For example, the waters off the coastal town of Oginohama is a key oyster farming site. Sanriku oyster farmers provide the offspring for nearly 80 percent of oysters farmed worldwide.[8] Now, commercial fishing is in danger of extinction. First, fishermen desperately need funds to rebuild their industry after most of their fishing boats were completely destroyed. In Miyagi Prefecture, waves destroyed 90 percent of the fishing vessels, that is, 12,000 out of 13,500 fishing boats, killed over 450 fishermen, and destroyed over 6,000 of their homes as well as flattening numerous fish processing plants.[9] Fishermen know well the telltale signs of a tsunami. They make their living on the sea's largesse combined with their skill and luck. For those whose daily bread comes from the ocean, stakes are high. Immediately after the 3.11 trembling, many fishermen headed in their boats to the open sea hoping to ride out the tsunami that would pass by underneath while it gained monster height the closer it came to shore. They had little time to spare, around twenty minutes. Some made it. Many did not. Next, with the threat of water being contaminated from the nuclear meltdown, the catch from their region, reputed to be some of the best in Japan, becomes less salable. Bad rumors, called *fūhyo-higai*, "rumor damage," has spread regarding contaminated fish and shellfish. Rumor mongering and stigma belongs to the so-called "fourth disaster," with lingering effects not only on the fishing industry, but on anyone who lives in and is associated with Fukushima.[10] Third, dark water from the tsunami filled villages with debris of all sorts. This was especially evident in Rikuzentakata where water carried buoys, tangled nets, ropes, cargo, wires, clothes, household items, and rotting bodies everywhere. Imagine being a professional diver tasked with disentangling debris.

Far-reaching damage was inflicted upon the fishing economy by 3.11. Add to this another, more human hurdle in terms of community's dark side that further hampers any possibility of restoration. In some places, there is an unwillingness on the part of surviving, older fishermen's "cooperatives" to open up fishing rights to outsiders. Extending fishing rights to these younger fishermen could help rekindle a collapsed industry. This closed-door mentality against outsiders regarding the opening of certain fishing zones, or *tokku*, certainly does not help matters.[11] Small-mindedness is often a rough and ready fabric of island or coastal living.

Jin's Struggle through Debris

Immediately after the tsunami, survivors were cut off from each other. The earthquake and waves killed all the power. Each person was in pure survival mode. And with the power down, connectivity—connecting with others through devices such as smartphones, iPads, Skype, email, and so forth—was impossible. Only later, in major cities like Sendai, power was restored in certain areas. And with that, connectivity was possible. Numerous survivors eventually connected with family and friends through Skyping, Facebook, Google Person Finder, and so forth. Facebook became a godsend for survivors to reconnect with each other and acquire locations of other surviving family members. Survivors were also able to offer mutual aid through opening up Twitter and Flickr accounts. Social media, in other words, was invaluable in enabling survivors to come into contact with others. Perhaps the best example of this was the YouTube message uploaded by Mayor Sakurai from his office to reach out to the world and plead for help. Nonetheless, in the days immediately following the enormous surge of sea on land, connecting with each other in this way was useless.

Jin Imai, in his mid-thirties, was at work with others when the quake struck. He knew right away from the giant quake that there was a strong likelihood the sea would soon follow. He was right. Within minutes, spotting the enormous wave coming onshore just north of Otsuchi, he and his coworkers, all good friends, were running for their lives up a hilly road. They had to reach higher ground. He even felt the water licking his heels as he sprinted as fast as he could. He finally made it to an elevation high enough to barely escape the hungry wave. His friends didn't make it. They were not quick enough. He did not see the wave swallow them. Looking back for just a split second meant life or death. He chose life. He did hear their muffled cries against the roaring locomotive sound of the sea behind him.

Jin escaped the hungry wave. He was now alone. One image propelled him uphill: faces of his wife and son. He *had to survive*, not for himself, but for them. He *had to survive* in order to find them and take care of them. They were his existence. He lived for them. And they needed him. All he saw when he raced up that hilly road were their faces, beckoning him to run as hard as he's ever ran. Jin is in good athletic shape, lean and fit. And now that he outraced the wave at his heels, he needed to outrace time and find his family. He was drenched, cold, muddied, on his own, and alone in the dark. At no other time in his life had this feeling of aloneness burrowed so deeply into him. With the power gone, his trusty mobile phone was of no use. Thoroughly exhausted, thoughts of finding his wife and son kept him awake, spurring him on to keep searching throughout the dark days and nights.

He waded through a soaked flatland filled with debris, dead bodies, and body parts, occasionally inspecting bodies that might resemble his family. Some of the dead were people he knew. The debris was lethal. The wave crushed everything, flattening wooden homes, office buildings, and shops, and gutting factories. The destruction left a wasteland of shattered glass, sharp metal and steel, wrecked vehicles, splintered planks, and rusty nails. One thought consumed him—to find his family. He wondered whether the waves had taken them. Fear shuts most of us down. For Jin, fear kept him awake. Amidst his fear, he believed deep in his marrow that they were still alive, and that they needed him.

For Jin's generation, digital devices were a natural part of the communication landscape. He and his peers all grew up thoroughly acquainted with the most advanced communication technologies. This was second-nature. He could easily navigate his way through smartphones, iPads, laptops, computers, computer hardware and software, connecting effortlessly through the various screens. But what good were they now? Texting, calling, emailing, Skyping, any kind of Internet access to his wife was now impossible. Connectivity in this sense was futile. He now had to navigate through mud, debris, and dead bodies. He had to find his family and bring them to safety. That was his sole driving thought. This was his reason, his "why" for surviving. And, as Nietzsche remarked, "If we have our own why in life, we shall get along with almost any how."[12] But this *how* seemed insurmountable. Jin had to somehow make his way through the water, mud, dangerous debris, raging fires, and bodies—bodies everywhere—to find his family.

He was not alone. Over the next two days, he spotted others who survived the waves, still in shock. They too desperately sought out their families, trudging through mud and debris, seeking signs of survival. For most of them, their search was futile. They never found their loved one or others. Finally, after the two longest days in his life, and searching through his crushed home, Jin found his wife and son safely huddled together with others in a makeshift shelter in town.

HAZARDS NATURAL AND HUMAN

All survivors faced this struggle through deadly debris. Indeed, Otsuchi represents all other towns along the coast in terms of the hazards survivors faced, the town's demography, poorly planned evacuation sites, and the monumental hurdles recovery workers encountered in their search for dead bodies. To being with, how do you survive without power? Without fresh water? Without heat and electricity? This is the plight millions face every day throughout our world. But Japan is not a Third World country. Yet this was now the challenge for those who made it through waves, collapsed buildings, and explosive fires. To add to this, during this rainy season at the tail end of a bitter winter, survival was all the more grueling, and for many unattainable.

Toilets were useless without a flushing mechanism. And keeping warm was crucial. Like our base camp for volunteers in Tono, kerosene heat replaced electricity. But kerosene was in short supply. And although it seemed natural enough to use wood as the principal source of fire and heat, almost all firewood was soaked through with sea and mud, and there was very little in the way of tools to cut down and chop up new wood from nearby forests. With tools destroyed, hand-built storage tanks had to be somehow put together to collect water. Like the Buddhist priest at the temple in Otsuchi who made his daily backbreaking pilgrimage climbing up and down the two high hills behind the temple to get fresh water, survivors often waited patiently in long lines carrying empty wooden buckets to fill them with water.

As for demography, towns along the Sanriku coast, for that matter throughout Japan, are increasingly populated by the aged and depopulated by the youth. As in numerous U.S. rural communities, many youth leave their rural roots to secure employment in metropolitan areas. Like many countries, Japan faces all sorts of prodigious problems from its rapidly aging population. As the disaster unfolded, numerous elderly were physically not mobile enough to get to an evacuation site in time to escape the onrushing waves. And mentally, they are naturally more prone to confusion during crises. For those that made it to designated sites, that did not guarantee survival. The powerful, underestimated tsunami destroyed over one hundred of these "safe zones," often at designated schools, community centers, and temples. Sophisticated earthquake warning systems surely alerted the public to an impending earthquake. Yet the unanticipated force of the magnitude 9 earthquake knocked out power lines and thereby annulled vital public tsunami warning systems.

While we had earlier described features of Japan's high-tech warning systems, let us elaborate further. For a country that once believed that either a giant catfish or a dragons' movements caused the earth's terrible and horrific trembling, or earthquakes, *jishin*, Japan has the world's most sophisticated earthquake early detection system. The system was officially launched rather recently on October 1, 2007. To prevent citizens from being caught off guard, Japan's elaborate system, built into mobile phones, issues signals of the initial P (primary) waves split seconds after eruption with, hopefully, just enough time to prepare in advance of the incoming and more destructive, lingering S (secondary) waves. This is just enough time for students to take cover under their school desks, drivers to pull off roads, trains to brake to prevent derailment, elevators to halt at the next floor, surgeons to end or postpone surgery, people to avoid glass and windows, and residents to shut off gas stoves.[13]

With the whole of Japan sitting on hundreds of faults, earthquakes favor no particular region. These faults are equal opportunity employers when it comes to earthquake alert and preparation. The Japan Meteorological Agency (JMA) oversees its earthquake early warning system. Once seismometers distributed throughout volatile areas send warning signals, monitoring stations immediately assess the need to issue a public warning through numerous media. Over one thousand seismographs are distributed throughout the country. These will emit information to the Japan Meteorological Agency Earthquake Phenomena Observation System which determines the eruption's epicenter and magnitude. This occurs within seconds of the eruption. Along with seismographs, seismic intensity meters further estimate the power and anticipated damage for a specific area. This is especially crucial in assessing the need to alert both disaster management professionals and workers and the general public.[14] Japan measures seismic intensity in *shindo* units from 0 to 7. 7 is the most intense and destructive. The March 11 magnitude 9 earthquake measured in at a seismic intensity of 7.

Based upon the scale of intensity, the JMA issues two types of warnings—advanced forecasts and alert warnings. Advanced forecasts are for those with advanced computer and alert systems who can then take immediate measures for their constituencies. These include school and hospital administrators, and top level decision makers at elevator facilities, railways, shopping centers, apartments, and construction sites. Alert warnings are for the public through media outlets like television, radio, and mobile devices as well as through outdoor loudspeakers.

Mobile phones are heavily used in Japan. When the 3.11 tragedy occurred, there were 119.5 million cell phone users out of a total population of 127.9 million.[15] On March 11, an advanced forecast was issued 5.4 seconds after detection of P-waves. 3.2 seconds later, warnings were issued to the public. Sendai, 129 kilometers (80 miles) west of the epicenter, is the affected region's largest city. The first warning was issued to the Sendai area just fifteen seconds before the arrival of S-waves and destructive shaking. Tokyo, 373 kilometers (231 miles) southwest of the epicenter, received warning sixty-five seconds before the six minute-long trembling.[16]

Seconds matter. Hence, the invaluable benefit of connectivity through devices. Alerts enabled facilities and trains to shut down. Due to East Japan Railways's sophisticated alert system, eleven Shinkansen trains operating in Tohoku stopped seconds prior to the trembling. Though the railway damage was extensive, this rapid alert and shutdown mitigated any further damage and no doubt saved countless lives. The 3.11 earthquake and ensuing tsunami, however, wielded a power that caused unimaginable destruction. The fault line spawning the earthquake at the epicenter stretched for hundreds of kilometers and covered such a distance that magnitude and seismic intensity were unprecedented. Because a major flaw in early warning systems is that they usually assume a single, more isolated source, the force of 3.11 was underestimated.[17] Ensuing power failures and chronic aftershocks added to 3.11's devastation.

Moreover, officially designated evacuation sites were often poorly placed, not on sufficiently high enough ground for a tsunami of this size, in which waves reached heights of ninety feet, for example in Minamisanriku and in Rikuzentakata. The seaside town of Rikuzentakata with its population of around 24,000 relied heavily on its tourist industry. For decades, it was one of the most heavily visited towns in Japan after workers planted nearly 70,000 stunning pine trees near the beaches. These pines trees were an impressive sight as townspeople and visitors would come to the beach and picnic under the cool shade of the trees. The seaside vista of beaches and pine trees was captivating. Having a high seawall would detract from the view. The breakwater that was constructed was therefore twenty-one feet high.

This turned out to be another fatal error in judgment. Waves from 58 to 90 feet in some places easily swept over the wall and flattened the town, which was literally "wiped off the map."[18] Waves over 90 feet reached the upper floors of the resort center by the beach. Though the pine trees helped break the tsunami's impact, all were crushed except for one, later known as the "miracle tree." After 3.11, this lone pine tree became the symbol of hope and resilience for the Japanese people. But, with time, even this "miracle tree" succumbed to rotting roots. Germinated seedlings were grafted onto the dying tree, and though much of the base of the tree is now contrived and artificially upheld, the tree still stands as a monument to rebirth. Citizens are not of one voice, however, regarding this miracle tree project. With an extreme scarcity of the town's monies and building resources, many people feel the monies spent on "restoring" the tree could have been better used. The cost of restoration was steep and could have been put to better use. Also, critics note that this effort to keep alive this symbol was more of an attempt on the part of government officials to convey a misleading image of new life, rebirth, and recovery to the world, an image that many local survivors have resented. For them, it is far from the truth. As living proof, tens of thousands of survivors still remain in cramped quarters in their temporary shelters (*ka-*

setsu jūtaku). In any case, calculations of the wave heights by the Japan Metereological Agency were way off base. The JMA was essentially unable to calculate the force of any earthquake over magnitude 8. Since the 3.11 magnitude was 9, estimated tsunami wave heights were low. For the Fukushima Daiichi nuclear plant, the JMA predicted wave heights of ten feet, a tragically low estimation given the ensuing ruination.

Search and Recovery

What about the obstacles recovery workers faced in their search for bodies? History and experience teach us an invaluable lesson. As described earlier, our neighbors nearby are usually the first on the scene, the first responders, in a disaster. This is often the case. Official responders—National Guard, Red Cross, FEMA—usually arrive later on the scene after citizens make an all-out effort loosely working together and doing what they can to help victims.

On the morning of December 6, 1917, when the Norwegian ship, *Imo*, collided into the *Mont Blanc*, a municipal ship that carried nearly three thousand tons of explosive powder, it set off a blast that shattered the port city of Halifax, Nova Scotia. A terrible fire spread killing over 1,500 people and destroying over a thousand buildings. The explosion has left an indelible scar in the memory of Halifax citizens.

My wife Brooke and I paid a number of visits to Halifax, and, each time, townspeople still talk about that dreadful day. In bookstores, books and pamphlets describe the unfolding events for the curious tourist. They especially point out the heroic efforts of Halifax residents in rescuing thousands of victims. One in particular, Samuel Henry Prince, an Anglican priest and curate of St. Paul's Church, set up his church as a sanctuary for over three hundred homeless victims. He went on to become a pioneer in disaster studies by writing a book on the Halifax tragedy. In it, he writes that in a disaster, "Life becomes like molten metal. It enters a state of flux from which it must reset upon a principle, a creed, or purpose."[19] And it is this principle of altruism and even self-sacrifice that drives ordinary people to rise to the occasion.

Tono citizens were the first to come to the scene of nearby stricken coastal towns like Otsuchi, Kamaishi, and Rikuzentakata. They became the lifeline for the survivors. Their acts reached heroic proportions, many of them risking their own safety to aid survivors. Official emergency response, on the other hand, takes time, coordinated communication, centralized command, adequate access and equipment, and jumping through the usual bureaucratic hoops. In the case of 3.11, however, one finds it hard to overstate the crucial role and lifesaving response of Japan's Self Defense Forces (SDF). The government sent the SDF immediately to the stricken coastline. Of Japan's nearly quarter-million SDF soldiers, over 100,000 of them were deployed. This disaster was much more powerful than the 1995 Kobe earthquake. After the Kobe quake, 20,000 troops were sent. However, because the tremendous impact of the 3.11 earthquake knocked out power, destroyed roads and transit systems including aircraft, travel to these stricken sites was nearly impossible. It took nearly a week before units of Japan's Self Defense Forces reached many of the towns after clearing away road debris, and barely making it over destroyed routes. Nonetheless, the SDF were the first official responders in the stricken towns. They conducted rescue and recovery efforts that far exceeded the thinkable. They enabled the rescue and recovery of nearly 19,000 victims, an amazing feat given the devastation they faced upon their later arrival in these towns. In order to recover bodies, which were everywhere, they had to navigate through the black mud filled with dangerous debris often hiding sharp metal, glass, and wood as well as containing toxic waste from numerous shattered factories.[20]

For SDF troops, theirs was a long, painful, and traumatic process of recovering countless bodies. These dead bodies are not viewed as simply corpses, but are often referred to as *hotoke*, or Buddhist spirits. They placed these *hotoke* in temporary morgues for family survivors to identify. Many civilians like Jin continued their search by visiting neighborhood community centers or schools where, on a slim chance, they might possibly recognize photos, photo albums, clothes, and household memorabilia collected and assorted by survivors, SDF, and volunteers. Volunteers manned tables where surviving families could inquire concerning the fate of their loved ones, many of whom were killed or missing. In towns such as Rikuzentakata where the tsunami destroyed all traces of official records of inhabitants, this was nearly impossible. Many survivors, in desiring to offer their help, also volunteered in taking shifts at these missing-persons tables. There was another type of table at these centers—reported death tables. Here, family members who were able to identify their loved ones' bodies filled out forms to make their official report and to arrange for temporary mass burial since many crematoriums were either destroyed or limited in the number of cremations they could perform per day. Those who were temporarily buried were eventually cremated.

When I returned in 2014, the mountains of debris had been cleared away. SDF troops, construction and bulldozing crews, and other workers accomplished this nearly impossible task. In the process, they also recovered photo albums, torn photos, and books. These were collected and sorted out in boxes and displayed at community centers for survivors to go through and hopefully piece together what might be left of their families. These are not just items, but memories, remnants of peoples' lives. Yet tons of debris still remain buried deep in the ground due to the force of the tsunami. Walking along the barren coastline, one can still find broken ceramic, pieces of tea cups, dishes, plates, kitchenware, household items, clothing, and other objects, all having their own personal history. With these immovable, stubborn relics of victims still in the hard ground, in ground that remains barren for hundreds of miles along the coast where houses once stood, is recovery even possible?

Proving Victimhood

Hurdles facing recovery are unnerving. Intuitively, it makes solid sense to begin reconstruction with rebuilding and then relocating destroyed homes to higher ground, further inland, away from the threatening sea. For example, the mayor of Rikuzentakata, Futoshi Toba, is strongly committed to bringing his beloved city back to life. Recently widowed, he raises his two sons. He has an eight-year plan to rebuild the city. Part of his plan consists in transforming the devastated and flattened areas by the beaches into an ocean park and resort. He is opposed to rebuilding on those same sites, formerly occupied by homes and shops. The rebuilding of apartments will be in elevated areas, with homes also projected to be built on higher ground. However, as we described earlier, who exactly owns the ruined land, homes, and properties? How will we determine this in order to obtain official approval of resettlement? The destruction of 3.11 not only wiped out homes, but virtually erased most city and local official records of home ownership and transactions. Official records were buried in mud and sucked out to sea. In most cases, official records of residence, ownership, addresses, family, and other tangible, official evidence of identity were completely wiped away, consumed by the waters, taken to sea, or churned into mud and soil. Therefore, for survivors, verifying their identity became a monumental hurdle in the disaster. Survivors have to prove they are victims in order to seek compensation from the local government, city or prefectural. For many, this is nearly impossible since official records are lost, co-mingled with debris, bodies and others' records.

Here we have a bureaucratic victimizing of victims, something similar to what happened following the Union Carbide chemical disaster in Bhopal, India. On December 2, 1984, the multinational company Union Carbide's pesticide plant leaked deadly gas including methyl isocynate throughout the area, killing over 8,000. Since then, at least another 8,000 have died from gas-related diseases. This remains the world's worst chemical disaster. The disaster was a result of incompetence, faulty facilities, lack of oversight and preparedness, mismanagement, and bungled politics. To add insult to injury, in order to benefit from government compensation, victims were required to have the proper certificate proving they were indeed victims. In order to have this certificate, they had to fill out numerous documents certifying residency, exposure, and so forth, which also meant having the requisite hospital admission papers with results of blood tests or x-rays. For the few who were able to procure such documents, they then had to fill out a form requesting to be recognized as a victim. Given the lower literacy levels of the poor inhabitants in the region, as environmental studies scholar S. Ravi Rajan describes:

> The complexity of the form, and the accompanying hassles of document procurement, made it difficult for the average Bhopal survivor to satisfy the requirements of the claims process. Ultimately, this meant that the victim became totally dependent on expensive touts [bribes] for an act as basic as filling out the claim.[21]

Months after Japan's 3.11 disaster, a popular poster circulated showing a group of *kasetsu* dwellers, and the poster was entitled "No looking back." In a way this makes good sense. But, sometimes it is only by looking back, for example, at family history and records, that one can move forward. Yet the sea has consumed family footprints so that remaining villagers stay entrapped in a long drawn-out, hopefully temporary, amnesia. Hence, this is one of the most daunting challenges in efforts at reconstruction. Who owns these properties? Where are they? How will we find out? Are there surviving families or relatives? Only then, once we establish ownership, whereabouts, and fate, can reconstruction move forward.

Along with this hurdle of ownership identity, the bottom line in whether or not recovery is possible comes down to resources. The region is certainly not without fiscal resources. On November 21, 2011, the Japanese government issued a sizeable fund, the Third Supplementary Budget, of 12 trillion yen (around $120 billion) especially targeted for reconstructing towns and homes, renewing the fishing industry, and building temporary shelters and offering monthly stipends.[22] Yet Tohoku is still plagued with insufficient resources and a diminishing economic infrastructure, mounting scarcity of jobs forcing youth to seek employment elsewhere, and a steady increase in the population of elderly. Furthermore, in Fukushima prefecture, clean-up along with decontamination will exact an enormous fiscal toll, let alone the task of rebuilding afterwards. There is the unimaginable task of scraping way the contaminated soil, six to twelve inches of the top ground layer, and then depositing it somewhere safe. Once the irradiated soil is scraped away, where then should it go? And how will workers replace contaminated soil with fertile soil from surrounding hills without incurring ecological scars? Add to all this the challenge in rebuilding schools and returning children to them. When schools were destroyed, there was often little choice but to move the children elsewhere to intact schools, often far away from their homes and communities, further uprooting the community for students.

But bottom-line resources are not only fiscal, but human. When confronting possible reasons for rebuilding—jobs, security, restoring homes and community—reality checks naturally compel us to raise questions as to whether there are grounds for hope, *kibō*. This question of human resources and grounds for hope takes us back to Jin and Bunzo.

HOPE LIES ONLY THROUGH CONNECTEDNESS

I first met Jin during my visit in January 2012. At that time, he had been leading young volunteers in helping to rebuild Otsuchi. They worked on various physical reconstruction projects as well as community building. As with Bunzo, social media and personal contacts soon enabled locals and people throughout Tohoku to know about him. People throughout the region, particularly youth, have signed up to work with him. By the time I reunited with Jin two years later, his group of volunteers had grown by the hundreds, most of them in high school and university.

Jin's message in guiding these young volunteers remains pretty straightforward. Building or restoring community, in the real sense, can only occur through interpersonal, face-to-face engagement with each other. It cannot occur simply through devices, devices that we have somehow come to think we desperately need and so much rely upon as our *only* way of connecting. Jin is untiring in stressing this. Over lunch together, he often repeated this to me with an intensity that reflects his measured thoughtfulness and personal commitment. And he walks the talk through his leadership and charism. Connecting devices, or connectivity, are, by all measures, vital in so many ways. But, from Jin's experience desperately seeking his family and navigating through debris-filled killing fields, he learned that these devices can never take the place of our flesh-and-blood being with and for each other, working, playing, crying, celebrating, and remembering together. As Watsuji puts it, we are indelibly in-between nature and all others. For Jin, Japan's only hope for the future lies in never forgetting this, in always remembering who we are. With this message in mind, Jin was also organizing and leading visits through ravaged areas along the coast for those many curious Japanese who wished to personally witness the extent of the destruction. This was his own self-made business. His "tours" were intended to impact upon "tourists" the long-lasting damage that townspeople along the coast now face.

Generations separate Jin Imai and Bunzo Kanayama, but they share a quiet bond in their love for their community. Jin had heard of the old truck driver and eventually met with him. They became fast friends and now collaborate. Jin spends part of his time driving volunteers to work for Bunzo. When I met again with Jin outside of Otsuchi in 2014, he and a busload of volunteers were preparing the hard ground to plant more *nanohana* along a windy barren stretch of land. It was exceptionally blustery that day, with sharp winds and freezing cold. Nonetheless, his group of around twenty worked hard at breaking and preparing the icy ground. Jin gave me a warm hug and greeting. After we talked for some time, he then introduced me to some of his workers. A few of them were from Tokyo. Others from Kyushu. And others from Hokkaido. Jin's message of hope through community continues to spread well beyond Tohoku. That day was one of the windiest, and we were all chilled to the bone. But our strong sense of connectedness kept us warm.

Children's Search and Rescue

That winter, I took a break from volunteering to visit my friend, Bunzo. Hiroko, in her usual gracious manner, kindly drove us to his new small living quarters in Otsuchi. Bunzo was there with two neighbors who also survived the tsunami. Their homes were destroyed as well. When I reunited with Bunzo, he gave me a giant hug and handshake. He was all excited about seeing me, and wanted to share with me his newfound friends—in Ireland. He knew my father was Irish-American, and he especially wanted to share this with me. Dad had died a few months earlier, just before Thanksgiving.

Some teachers from Northern Ireland's Carrickfergus Model Primary School and the Japan Society of Northern Ireland had friends who visited Japan after the disaster. They had heard from their friends about Bunzo's inspiring work and his *nanohana* festival. They wanted to see for themselves. Their young students were learning about Japan, its history, culture, and language. So the teachers visited Bunzo during his festival. Afterwards, they returned to Ireland and shared stories of their visit with their students. Each student then wrote his and her own small, precious note of thanks, well-wishes, good luck, and prayers to Bunzo, often in Japanese, in the design of a yellow flower. The teachers assembled these notes into albums and sent them to Bunzo, along with class photos. There's our inescapable in-between at work again, this time connecting Japan and Ireland. Bunzo, his friends, Hiroko and I spent a good part of that afternoon at his kitchen table looking through the albums and talking about Ireland and where my father's father and mother lived. See figure 5.1 for a photo of Bunzo and Hiroko with the school album from Northern Ireland. Figure 5.2 shows notes from children in Carrickfergus Primary School.

Aside from my Irish-American father marrying my Japanese mother, I now believe there is more to the link between Japan and Ireland than meets the eye. I came upon a bit of Ireland in Tono. Toshihiro and Yuko Kuramoto run the wonderful "Bed and Breakfast Kuranoya" on the outskirts of Tono, a spotless and comfortable blend of traditional and modern Japan. As a needed respite from my stay at the volunteer quarters, I spent the night here on St.

Figure 5.1. Bunzo and Hiroko with album from Northern Ireland

Patrick's Day, my dad's biggest day. When I first entered the inn, there were soft sounds of traditional Irish music. For the innkeepers this was not intended in celebration of "the day." In fact, when I mentioned this to Toshi and Yuko, they had no idea of the significance of March 17. Toshi and Yuko simply loved Celtic music and often dream of visiting Ireland. They were drawn to Ireland by its music, beautiful photos of green and lush mountains blended with harsh and barren hills, and by its legends and magical stories.

They found something similar in Tono. The couple moved to Tono from Tokyo, where Toshi worked for Japan Newsweek. When they visited Tono a few years back, they fell in love with its tranquility and peace along with its ruggedness. We talked about the various similarities between Ireland and Tono: their folktales, strong community-mindedness, love of tradition and its music, leprechauns and the mischievous *kappa*, people's closeness to nature and their love of mountains, even in harsh weather always followed by abundant green.

This was all auspicious. Nothing occurs by accident. Before my trip to Japan, in New York, I had shopped around for the right gift for Bunzo. When I gave it to him, his eyes widened. It was an Irish shamrock, encased in crystal. This was propitious in another sense. In Bunzo's case, we have

Figure 5.2. Students' notes to Bunzo from Carrickfergus Primary School

schoolchildren in Ireland sending him their photos. And throughout Japan for nearly a year after 3.11, elementary schoolchildren were immersed in cleaning photos that SDF troops and others had retrieved from the stricken coastal towns. These were single photos and albums of the dead and missing that were sent to schools. Japanese schoolchildren carefully, meticulously removed dirt and mud, one piece, one spot, one section at a time. Schools then returned these cleaned photos and albums back to towns all along the Sanriku coast. These salvaged photos of young children, families, grandparents, and pets are especially haunting and remain faint traces of an incomplete, interrupted life, mere whispers of an unfinished conversation.

When our particularity, our sense of who we uniquely are, is threatened, we strive all the more to be especially meticulous in the particulars, in the details. Envision hundreds of thousands of children throughout the country carefully cleaning muddied photos with toothbrushes, cotton swabs, and wipies, gently rubbing and drying. This was their own search and rescue effort, via photos and other *omoide no mono*, "memory items," outsourced to children, who naturally embody the future, and who help save the past. In the aftermath of 3.11, the entire nation is involved in this piecemeal rescue effort to preserve memory. In most cases, memories are all that is left. After grouping all these together in albums, volunteers and those who are left of the community then place theses photos and memories into boxes and bring them with their albums to selected municipal centers. Here, survivors would regularly come to conduct their own search and recovery and bring their memory items to their final destination, home. In the corridors of death, hope dies last, if it dies at all.

NOTES

1. See video at Manish Prabhune, "Rapeseed blossoms to start off the spring season," *Japan Travel.com*, March 2013, accessed Feb. 13, 2015, http://en.japantravel.com/view/yellow-flowers-at-hamarikyu-garden.

2. Hermann Hesse, "Stages," in *The Glass Bead Game* (*Magister Ludi*), trans. Richard and Clara Winston (New York: Owl Books, Henry Holt, 1990), 444.

3. Kenji Miyazawa, "The Morning of Last Farewell," in Kenji Miyazawa, *Strong in the Rain: Selected Poems*, trans. Roger Pulvers (Northumberland, UK: Bloodaxe Books, 2007), 79.

4. Kenji Miyazawa, "Bare Feet of Light," cited in Kenji Miyazawa, *Strong in the Rain*, trans. Roger Pulvers, 24.

5. Rob Gilhooly, "Fukushima man opts to be guinea pig," *Japan Times*, Sept. 14, 2011, accessed January 2, 2015, http://www.japantimes.co.jp/news/2011/09/14/national/fukushima-man-opts-to-be-guinea-pig/#.VOE0OWctDcs.

6. Lucy Birmingham and David McNeil, *Strong in the Rain: Surviving Japan's Earthquake, Tsunami, and Fukushima Nuclear Disaster* (New York: Palgrave Macmillan, 2012), 182.

7. Anne Allison, *Precarious Japan* (Durham, NC: Duke University Press, 2013), 198.

8. Birmingham and McNeil, *Strong in the Rain*, 32.

9. Johannes Wilhelm and Alyne Delaney, "No Homes, No Boats, No Rafts: Miyagi Coastal People in the Aftermath of Disaster," in *Japan Copes with Calamity: Ethnographies of the Earthquake, Tsunami and Nuclear Disasters of March 2011*, 2nd ed., edited by Tom Gill, Brigitte Steger, and David H. Slater (Bern, Switzerland: Peter Lang, 2013, 2014), 103.

10. Wilhelm and Delaney, "No Homes," 104.

11. Wilhelm and Delaney, "No Homes," 108.

12. Friedrich Nietzsche, "Twilight of the Idols," trans. Walter Kaufmann, in *The Portable Nietzsche* (New York: Viking Press, 1968), accessed January 27, 2015, http://www.inp.uw.edu.pl/mdsie/Political_Thought/twilight-of-the-idols-friedrich-neitzsche.pdf.

13. Erika Yamasaki provides one of the clearest accounts of this elaborate system. See Erika Yamasaki, "What We Can Learn from Japan's Early Earthquake Warning System," *Momemtum* Vol. 1, Issue 1 (April 18, 2012): 9; accessed Sept. 1, 2014, http://repository.upenn.edu/cgi/viewcontent.cgi?article=1022&context=momentum.

14. Yamasaki, "What We Can Learn," 4.

15. Yamasaki, "What We Can Learn," 9, note 1.

16. Yamasaki, "What We Can Learn," 10.

17. Yamasaki, "What We Can Learn," 12.

18. Birmingham and McNeil, *Strong in the Rain*, 55.

19. Samuel Henry Prince, *Catastrophe and Social Change: Based upon a Sociological Study of the Halifax Disaster* (New York: Columbia University Press, 1920), 16, cited in Rebecca Solnit, *A Paradise Built in Hell: The Extraordinary Communities That Arise in Disaster* (New York: Penguin Books, 2009), 79.

20. Birmingham and McNeil, *Strong in the Rain*, 126.

21. S. Ravi Rajan, "Missing Expertise, Categorical Politics, and Chronic Disasters: The Case of Bhopal," in *Catastrophe & Culture: The Anthropology of Disaster*, ed. Susanna M. Hoffman and Anthony Oliver-Smith (Santa Fe, NM: School of American Research Press, 2002), 246.

22. Wilhelm and Delaney, "No Homes," 109.

Chapter Six

Volunteering—Cold, Snow, Rain, Wind, and Mud

The cold started at Shin-Hanamaki station, just before I boarded the Kamaishi Line to Tono. It was as if the large, foreboding figures of mountain gods, or *shishi*, in the station, with their wild, black faces and gold teeth were warning me, "You who come here, be prepared." Eager to start my volunteer stint, I left the warm comfort of the shinkansen from Tokyo with my luggage weighing over 55 lbs. Shin-Hanamaki station had no elevator. My bad back and recent triple hernia told me that my suitcase was far too heavy. *Omoi desu*. Especially climbing down over forty steps and then climbing back up another forty steps to board the local train to Kamaishi. I felt the strain in my lower abdomen. Nevertheless, preparing for the cold that others had warned me about, I had extra base layers and outerwear, fortunately, as this turned out to be the coldest sustained period I've ever faced. Nearly every day from then on, both wind and cold cut to the bone. For that month of March, it never left. It stayed with us—bitter cold, wind, and snow.

The faces on the Kamaishi Line told a thousand stories. They were old, young, and in-between, mostly old with lines carved into their faces from a lifetime of hard work and struggle. Theirs was a shared struggle, not solitary. Usually, in Tokyo or some other major Japanese city, I get occasional glances from train passengers. They know I'm different. They quickly look away when I return their look. But not here, not on this line. Only a few school boys in their uniforms. And what caught my attention was that these students were actually reading books, school books. None of them were absorbed in smart phones, iPads, or mobile phones. None were peering into some private screen-world. Their world in Tohoku was real enough. As we neared Tono, I knew I was entering an entirely different dimension from bustling Tokyo. The snow steadily mounted, adding to the separation. And passengers' faces, worn-down with creases of struggle, yet gentle, all tell a story of tough survival.

Hiroko met me at the Tono City station with her ever-present warm and kind smile. She had that same serene look when I met her two years ago in January 2012, just ten months after 3.11. Yet, there was more concern etched in her eyes and face. She led me to her office right across the street from Tono station. She now worked as a manager for the Tono Travel Agency, and the office was spacious, welcoming, and warm. After helping me navigate the ritual of obtaining volunteer insurance, buying rubber boots, gloves, and food and delivering me to Magokoro Net's Tono headquarters on the outskirts of Tono City, I finally settled in, ready to start my work early next morning.

HEADQUARTERS

Magokoro Net's headquarters, where we stayed, was in a makeshift building. It was, at best, thinly insulated, but good enough to protect us from Tono's especially harsh winds this time of year. Though Spartan, it was comfortable enough. Men and women had their separate sleeping quarters, with traditional tatami mats lining the floors. The rooms were bare and cold. The men's quarters were closer to the front entrance and near the group washroom that separated us from the icy cold with sheets of plastic and a thin but sufficient barrier from the outside.

Occasionally some pipes would freeze overnight when one of us left the washroom's main waterline open. For our building to comfortably accommodate us, anywhere from ten to fifty, we had four working sinks, six toilets, one small kitchen, two washers and dryers. This was Tono's coldest time of year with its strong winds, incessant heavy snow, and stubborn ice, giving us a slight taste of the weather conditions survivors faced in the wake of 3.11. Our building had plastic sheeting on the interior to keep out cold drafts. In this dead of winter, we did not use electricity for heating, but instead relied on a few kerosene heaters: one in each sleeping room (men and women), and one in the meeting room. We were sparing in our use of kerosene and electric lighting in a collective effort, nationwide, to conserve power. At night, we would either turn down or shut off the lone kerosene heater where we slept. And in the morning . . . well, nothing like an early morning cold water wash to stir the senses. Yet compared to the evacuation shelters cramped with survivors, our refuge was sheer luxury.

That first night was one of the coldest. I barely slept two hours. And the cold never left. When it became even colder my second night, I seriously wondered whether I would make it through the weeks ahead. Nonetheless, each night I buried myself in my sleeping bag, wearing base layers with my New England Patriots' woolen cap pulled over my ears, watching my breath form dark clouds, and preparing myself for what became some of the most memorable moments in my life. Though definitely my longest, sustained period of cold, the many warm hearts of volunteers, staff, friends, and 3.11 survivors provided more than enough warmth.

Recall how Watsuji Tetsuro writes in his *Fudo* that our climate indelibly impacts the ways we look at life, what we value, what we fear, what most matters. We cannot escape the weather. We are all pulled by it, ruled by its temperament. Weather determines what we wear, eat, drink, how we sleep, and where and how we choose to live. Moreover, our natural climate, our weather, ever-present and defining, always triumphs over our paltry, human efforts to subdue Nature. This was clearly the case with our cold Tohoku and Sanriku coast winter. Yet as Watsuji later elaborated in his *Rinrigaku*, climate is not simply our natural climate, our weather. Weather does not always claim victory. For climate includes social climate—our human bond and collective struggle to survive.

This is the region's harshest time of the year, with temperatures often falling below zero and, combined with razor sharp winds in valleys like Tono surrounded by mountains, incessant snow and rain prolonged a bitter, icy chill. Yet there lies an inseparable gap between what we faced and what survivors experienced. We volunteers were prepared for this. I always wore my base layers, cap, and gloves. We chose this. Most survivors and victims had lost their clothes. Clothes were later offered to them by villagers in Tono and elsewhere. We volunteers had a roof with at least minimal insulation, lighting, and kerosene heaters. Survivors were jammed into wintry shelters without power, running water, and little heat. For them, harsh weather added insult to their insufferable injury. Having survived, they had to endure bitter, bone-cutting cold, wet, and wind.

Why was I here? Born in Fukuoka, Kyushu, as Kenji Kimura, having a Japanese mother and Irish-American father, I became especially drawn to matters of Japanese culture, the Irish, other cultures, and how different cultures understand and communicate with each other. Nonetheless, to many Japanese, I'm still *gaijin*, a foreigner. Volunteers would often ask me, "Why are you here?" When Japan's 3.11 calamity unfolded, I silently promised my mother Misae, who died in December 2009, to learn more about the disaster and share this with others. I wanted to better understand and appreciate the complexities of the tragedy, its aftermath, and impact by working more closely in stricken areas in Tohoku. My three principal contacts during my earlier visit in 2012—my friends Kunio Aoki, manager of the East Japan Railway Culture Foundation, his assistant Mizuki Naito Abe, and Hiroko

Suzuki, who now worked at Tono Travel Agency—encouraged me to consider volunteer work. Hiroko suggested that I join the local group Tono Magokoro (*magokoro* literally means "true heart") Network.

This experience on the ground remains for me one of the most enriching learning experiences in my life. This was why I wanted to be here in Tohoku during this season. While I could never fathom or experience the terrible suffering of those countless victims of 3.11, I needed to gain some slight sense of what they encountered, weather-wise, to get inside their skin, even though it was a mere fraction of their immeasurable hardship. To some readers, I'm sure this seems rather naïve, even reckless. Nonetheless, I needed to experience at least a thin taste of the harsh climate survivors faced, even if miniscule.

Tono Magokoro Network

Humble. That's the word. The volunteers and kind Magokoro Net staff do not think of themselves as doing anything special or heroic. Every staff member I met played down their own particularities and, instead, insisted they were simply doing what others would do if they had the opportunity. Each person was humble in the genuine sense. Many of the volunteers at this time of year were university students. University students were having their winter break, or "Spring break" as we call it in the United States. However, whereas American students have usually a week off, Japanese students have two months. For these Japanese students, they had the time, and they chose to spend a part of it this way. For each of them, they could not help but be of service. They simply chose to do what had to be done, without fanfare. Figure 6.1 shows me surrounded by some student volunteers. Figure 6.2 shows other volunteers with their leader Akira Kamatsu in front.

Now for a bit of background of Tono Magokoro Net.[1] Its Public Affairs Manager, Makoto Yanagisawa, generously offered me details of the organization's history. He taught earlier at the Medical College of Georgia, but, after 3.11, left for Tono to work towards providing aid to survivors.[2] Stanford University doctoral fellow and former volunteer Peggy Fan also offers a concise report on the work of the organization.[3] So does Tom Wilson in his Oct. 26, 2011, article in *Time*.[4]

Tono Magokoro Net is exactly that, a unique volunteer network that joins together five Tono citizen groups and Tono's social welfare councils. Prior to the disaster, it was historically unprecedented for social welfare councils to work together with citizen groups. Immediately after 3.11, they each worked independently to provide relief to tsunami victims in Iwate coastal regions. However, the magnitude and scope of the tsunami destruction exceeded all imagination. This was a worst case scenario they had not planned for. So, they put their efforts together, officially forming Tono Magokoro Net on

Figure 6.1. Author with student volunteers at Magokoro headquarters

March 28th, 2011. Magokoro Net headquarters started with humble origins in a hallway at Tono Sogo Fukushi Center (Social Welfare Center). It later moved to a temporary prefabricated building in Tono Joka Center (Sewage Treatment Plant). Its current site is on the outskirts of Tono City, in a valley surrounded by small to moderate-sized mountains. Steady winds, from gentle to severe, exposed us to the wind chill that made the headquarters particularly cold during these months. The Net has now grown in membership and has a branch office in Kanda, Tokyo.[5]

Not all volunteer organizations operated effectively. Nothing is fail-proof. To begin with, there was no centralized government coordinating system for work with NGOs (Non-Governmental Organizations) and nonprofit groups who wished to bring aid to stricken areas. This made it difficult for volunteers to work effectively as they first needed to contact previously established volunteer centers. Yet many of these centers were unable to work with local administrators and city officials since the latter lacked direct guidance from prefectural governments, and many town officials were killed or missing.[6] Next, there were often disconnects between volunteer centers and local realities. For instance, one center attempted to reach out to local citizens, trying to find out who needed help and in what ways, by way of the Internet. But

Figure 6.2. Volunteers with leader Akira Kamatsu, Magokoro headquarters

there was simply no Internet. The events of 3.11 disabled Internet use for many inhabitants. Moreover, many inhabitants are elderly, and many elderly do not use the Internet and are unable to navigate anything online.[7] However, because Magokoro Net was based in Tono, and, as described earlier, Tono was operating more systematically and efficiently under the strong leadership of its mayor, Toshiaki Honda, the Net consistently offered vital service to the coastal communities.

The Net's mission is simple—to provide multiple needs for the tsunami victims. At first, this meant offering necessary food, water, supplies, clothes, umbrellas, boots, and gloves. This also included removing debris. Debris in all forms—splintered wood, shattered homes, twisted wreckage from cars, boats, sharp glass and metal, paper, plastic, shredded clothes, and so forth—covered the landscape. Months later, mountains of debris covered the devastated areas like anthills. The Net now is especially committed to serve the countless survivors who live in temporary shelters. The word "temporary" is ironic. To this day, over four years later, hundreds of thousands of survivors inhabit these *kasetsu jūtaku* (temporary houses) and, given the local and national governments' snail pace of progress, will most likely live here for at least another decade. And, especially since tsunami victims now need "houses" (places they can call "home") and jobs, the Net focuses on rebuilding community and creating jobs, particularly on farms in Otsuchi.

Ryoichi Usuzawa is one of these survivors, a dignified gentleman in his mid-sixties with a handsome, sculpted face. He regularly shares with volunteers his own experiences as a tsunami victim. The tsunami engulfed and destroyed his and neighbors' homes in Otsuchi. Soon after the trembling subsided from the earthquake, against his wife and daughter-in-law's pleas to leave their home and escape the oncoming tsunami, he insisted the waves would not reach their home. They never had in the past. When the wall of water came roaring towards them, his wife, daughter-in-law, and eleven-month-old grandson immediately headed to the evacuation site on higher ground while he headed upstairs to save their dog Taro. By the time he started to come down the stairs with Taro, it was too late. The first floor was entirely filled to the ceiling. His only option was to go to the roof. From there, he stared in disbelief. The entire area was buried in dark, moving water. "It was like a huge washing machine was spinning the whole town . . . everything moving counter-clockwise."[8] To make matters worse, waves violently tossed cars about with their blaring horns, many of these into convenience stores supplied with propane gas tanks. Everything was exploding as a horrifying inferno was spreading towards him.

It was a miracle he and Taro survived. By grabbing the dead power cable above his head, dead since the tremblor knocked out all power, he slid carefully over the waters until the cable led to a concrete building. There, as the first wave receded, he felt safe. But there was a second wave, this one worse. As the waters rose up to his neck, he thought it was over. Just when the waters inched higher and he and Taro could now only breathe into gaps in the ceiling, the waters stopped, and ever so slowly descended. It was serendipitous. There was a firefighter close by who heard his cries for help. Chiaki Michibata helped Ryoichi and Taro make it safely to dry ground. Together, they went to the evacuation center, and Ryoichi reunited with his family.

Ryoichi believes this was not luck but the work of God: the overhead cable leading to the one concrete building in the area, a voice telling him to hang on to the cable, and Chiaki. "I believe that God let me live. What can I do with this life saved by God? . . . The only thing I can do is hold hands with people who are suffering . . . and eventually rebuild the town."[9] Ryoichi and his wife and their Taro were then placed in a *kasetsu* in Otsuchi where they still live. Ryoichi is also one of the directors of Magokoro Net. He spends much of his time traveling throughout the country speaking to groups about the tragedy, the devastating impact on the environment, his own experience as a shelter survivor, and the plight of other survivors.

Exceptionally tall for a Japanese, Ryoichi is especially humble and very interested in my reasons for working there. His eyes tell it all. They penetrate into your soul and convey his remarkable charisma. After he delivered a moving account of the events of 3.11 to our group of volunteers, he and I met. He told me more about his ordeal as a survivor living alone in a temporary shelter. Many survivors often wonder whether they were fortunate to survive. They feel isolated, abandoned, without family, friends, neighbors, and thoroughly disconnected. They are shadow people, living on the edge. Many whom he knew contemplated suicide. He believes there is a reason why he was saved. This is his reason to live. He aims to help other survivors, share his stories, and to cultivate in their marrow an appreciation for being alive, and most importantly, to never lose hope. Ryoichi and I made a pact to reunite when I return. He has invited me to share tea, stories, and presence.

Throughout my stint, I was blessed with committed leaders. Keita Inoue probably knows more about the disaster than a library of books. With his shaved head, shining eyes, and steady smile, he reminds me of a lighthearted Zen Buddhist monk. Keita's small stature is deceiving. He is healthy, strong, robust, and physically in top shape. He arrives early every morning to lead us through the morning exercises called *rajio taiso*. *Rajio taiso*, or "radio exercises," is popular throughout Japan. Workers, school children, elderly, people everywhere do their regular exercise morning routine of calisthenics accompanied by music. Keita has perpetual energy, and even after a long working day he will talk far into the evening with volunteers. On our daily ride to work at some coastal town—usually Otsuchi, Kamaishi, or Rikuzentakata—Keita would talk to us about the area, its history and people, and how 3.11 and its aftermath has affected the area. Another leader is the quiet, confident, stern-looking Tetsuya Arakawa, always with his boots and standing at attention. He usually led the group of us who worked at the Net's annex outside of Otsuchi. Our work there consisted in transforming the property into a useful community center for nearby *kasetsu* residents, a place for them to gather together for special events or festivals. We would also clear out an area of farmable land where they could plant their own vegetables and flowers. Tetsuya was reserved, yet immensely generous. He occasionally treated us to grilled fish during our lunch break. And then there is Akira Kamatsu, never without a smile and helpful hand. During those months, Akira usually led the group that worked at a Magokoro center where persons who were disabled volunteered their services by designing cards, cutting posters, putting together packets, stuffing envelopes, all with the utmost attentiveness and precision.

To Volunteer Is to Acknowledge

When we've reached the point whereby another's suffering means little or nothing to us, we've further erased any possibility of redemption—a suicide of our soul. For me, and I believe for many of us volunteers in Tohoku, our work was a sort of gateway to healing. By healing I mean transforming the mere intellectual knowing about 3.11 and its wake of suffering and lingering aftermath and, in the case of radiation, the terrible anxiety over what is yet to come, into an "acknowledging," to use Stanley Cohen's incisive distinction, of that horror. To acknowledge is the bread of conversion. To acknowledge means we internalize what we know in ways that transform us. It requires opening our eyes so that we really see.

Normally, what we choose to see is rear-view reality, rear-view suffering. This more easily fits into our line of vision. We notice it as we pass by it. It is usually the case that merely knowing on a cognitive level, like wearing T-shirts with *Gaman Rikuzentakaka* and *Gambatte Nippon* and adopting slogans, is about as far as we go. We remain accidental activists. As Cohen puts it, we still play in the universal board game of Denial.[10] To acknowledge transforms in a way that moves us out of our conventional zone of silence, that zone which easily spawns indifference. To acknowledge moves us to act on the grounds of what we know.

Such is what we shared, our common ground, as volunteers, particularly for me as an academic. Academic colleagues have asked me, "So, what did you do on your sabbatical? What was your research? What did you read?" Many stared back quizzically when I responded, "I read little. I worked as a volunteer with Tono Magokoro Net, spending cold, windy, and rainy days shoveling snow and mud, moving rocks, splitting wood, cleaning rooms and windows, setting up temporary platforms. All in all, working most of the time in snow, mud, and debris." I was not caged in a book written by others *about* survivors and the ubiquitous *kasetsu*, nor isolated in my office reflecting on third-hand accounts. Rather than weave around me that safe net of cognitive empathy, I wished to *see for myself* the current plight of survivors and feel their circumstances on the ground, not mediated solely through distanced narratives from academics and others who maintain their safety zones, those whom some critics call "parachute journalists" who profess expertise with little understanding of the terrain. We can say the same of "parachute philosophers" (like myself perhaps to a degree) who utilize the disinherited via detached research, in a sense further marginalizing those on the periphery.[11]

Unlike many Japanese, especially those who live in major cities like Tokyo and Osaka, volunteers choose to unwrap their everyday protective gauze. This is the illusion that because it's been over three years now since 3.11, all is now back to normal. The volunteers I worked with seem to cry out for the need to *see more clearly*. And seeing more clearly does not mean taking photos of the devastation. There were strict rules against unsolicited photo-taking. Yet these young people didn't need to be told the rules. They intuitively sensed the profanity of thoughtlessly snapping photos. They seemed to naturally know in their hearts that the reality of such suffering transcends the mundane. No photo can ever come close to capturing the anguish that echoed daily around us as we worked in a massive, barren amphitheater of silence. Almost all volunteers, most of them university students, were not compelled to capture the experience of being in these devastated areas with photos, on some screen. This is quite different from the screen life many young people now inhabit. Our screen lives have sadly become our vocabulary of ordering, of our day-to-day routine. We, however, were literally working in fields of death. These fields were one mass grave site, a site of mass destruction. I could still hear their stifled screams.

These young volunteers in particular showed a quiet and prayerful attitude at the various shrines commemorating the disaster and the victims. These are the moments when language is insignificant. When we avoid having to focus on words. When we live in and speak simply through silence. These are the cracks in our regular discourse that say more than any words.

Schedule

No two days were alike at Magokoro Net. Still, we held to a strict schedule. I usually slipped out of my sleeping bag at 5:22 a.m. (my body was set to that time) to use the wash room before others woke up. I was immediately smacked with the chill of our sleeping hall. Our only source of heat, one kerosene heater, was either turned off or set very low to conserve fuel overnight. Tiptoeing into the wash room, careful not to step on anyone in their sleeping bag, the ice cold air shocked me into full awakening. Our washroom had four working sinks, and occasionally the pipes froze overnight, leaving us with only cold water in the morning. I would splash freezing water over my face and clean up and waken the senses. After breakfast, my usual fare of orange juice, tea, yogurt, bread, and fruit, we then assembled for *rajio taiso*. I've come to know *rajio taiso* by heart—its routine, words, and uplifting music. Keita led us in our exercises, after which we lined up, according to our choice, to work in various coastal towns. After instructions from Keita and Tetsuya, we boarded vans to our respective destinations. The staff assembled outside in the cold and snow to wave us off. This send-off was the usual ritual. Whenever we departed as a group, staff would give us the gracious farewell to express good luck and gratitude.

Our drive down the winding paths to the coast took around forty minutes. We drove through small towns, most of them in the process of rebuilding. Upon reaching our destination, we faced still-ravaged and flattened areas, spotted with tractors and digging equipment. Three years after the impact, I expected more solid evidence of reconstruction. But although most of the debris had been removed, a monumental effort by any measure, numerous areas still remain barren. Upon arrival and further instructions, we worked until noon.

For me, and I believe for most of the volunteers, working in coastal towns obliterated by the tsunami was a profoundly spiritual experience. This was particularly felt the closer we worked to the shoreline, where the sea meets land. This shoreline represents a sort of mystical border, separating the realm of the unknown and the known. In the case of 3.11, the sea violently crossed that customary border and shredded entire towns, swallowed inhabitants, pulled them back into its belly, and spat many back out. The barren land, once occupied by homes that have been shredded by the waves, were for us sacred land. Though barren, the spirits of the victims still seem to inhabit the area. Theirs is an unmistakable presence. I hear them silently pleading with us, the living, not to forget them, not to forget our living together, to remember our inseparableness. The soil was sacred, holy, and not to be defiled by the distanced peers of tourists and the indifferent gaze of spectators with their iPods and cameras.

Along with a handful of us, most of my duties consisted of heavy work (*chikara shigoto*), chopping ice, shoveling snow and mud, lifting and moving rocks to make spaces to grow community flower and vegetable gardens in various *kasetsu* when Spring comes, splitting firewood, moving appliances (stoves and refrigerators) in *kasetsu* units that were to be used for community get-togethers.

From noon to 1 p.m., we lunched, often at *kasetsu* eateries which desperately needed business. These were temporary makeshift facilities run by survivors and victims' families. At one *kasetsu* seafood restaurant in Rikuzentakata, the owner had lost family and neighbors to the tsunami, where waves reached heights of nineteen meters, or sixty-two feet. Knowing ahead of time that we would be coming there to eat, he put together for us a slide show of the town's colorful history with its rich traditions, beauty and allure as a seaside resort, followed by scenes of the disaster that ravaged the entire town. Now, three years later, Rikuzentakata, like many coastal towns, remains flattened and showing only slight signs of rebuilding.

After lunch, we then worked from one to three in the afternoon and then headed back to base camp. Upon returning to our Magokoro quarters, we diligently worked together, everyone without exception, cleaning the facility—floors, kitchen, sleeping quarters, washrooms, toilets, meeting areas, tatami mats, and so forth. Cleaning took on a special quality. It was more meaningful than meets the eye. Following a day working and sweating in disaster sites where a black sea swallowed up any living being it could claim, and returning to Magokoro Net, our first task was cleaning. It was, in a sense, our way of recovering as a group, *han*, to clean and purify ourselves from our contact with death. Cleaning together was a way of regaining equilibrium, normalcy, and order. Magokoro Net had become our collective, temporary *uchi*, our "home" away from home. And cleaning affirmed this *uchi*, safe and protected from outside dirt, dust, mud, snow, disorder, and violent unpredictability. Japanese studies scholar Brigitte Steger artfully writes about this need for order through cleanliness.[12]

At five in the afternoon, we held our group meeting. Here, leaders would share their insights, introduce new arrivals, and give reports. Volunteers would also share with us their personal perspectives from their work experience and encounters. Also, volunteers who were leaving—volunteers usually stayed for around three days—offered lessons they'd gleaned and gratitude to all for enabling them to have this experience. Throughout all this, I was impressed by students' articulateness in why they came to Tohoku to volunteer, what they discovered and learned along the way, and what they hope to share with others when they return to the comfort of their homes. Almost all who leave pledge to return. As Akiko often repeated, "There is no recovery until everyone is recovered, no exceptions."

Staff members tell me that, of foreigners volunteering during winter, I hold the record for the length of time. I think they were just trying to make me feel good, to encourage me. To be sure, I gave many farewells to those returning to their homes, and offered as many welcomes to new recruits early next morning. At these meetings, it was abundantly clear that these young people had kind hearts, filled with empathy and compassion. Despite all the tragedy and sadness we've encountered throughout our work, if they represent the future of Japan, then there is much promise and hope for my homeland. Our meetings usually lasted an hour. After that, dinner. Some of us later trekked to the town hotel to have a bath, Japanese style and a type of heaven. Others clustered in small groups, warming themselves around the kerosene heater in our meeting room. Centered around the warmth, rubbing our hands by the heater, chatting small talk, was our way of bonding after working together in the cold, bearing witness to *kasetsu* dwellers' fear and future uncertainty. Hanging out together around the heater reaffirmed our natural interdependence, our in-betweenness. At ten at night, it was lights out before our new day ahead.

SNAPSHOTS

Our work was rather ordinary. Nothing remarkable or special. We worked in snow, rain, and mud. There is nothing more ordinary than mud. The mud here took on a certain meaning. What were once solid homes are now crushed, liquefied, pulverized and buried in mud. Waves have churned soil into a dark, heavy goo that grabs everything it meets including debris, clothes, shoes, tea-cups, saucers, and bodies, and innumerable bodies—livestock, pets, humans. Our volunteer work was all about this "getting into the mud" together, to lift one shovelful at a time, no doubt tempering us from any need for immediate gratification. Our getting into the mud is a way of somehow stepping into the pain of others. Working in the mud, in the ordinary, brings with it a way of noticing the small things, seemingly insignificant, that take on a special quality, the little snapshots we remember and that, put together, make the story. In the long run, we define our lives by small acts. So also, these little stories make all the difference.

Again, each day was different. It depended upon the weather and specific needs in the numerous shelters. For me, following a hard overnight snowfall, I often spent the day shoveling heavy snow at *kasetsu* in Otsuchi and Kamaishi. Freezing overnight temperatures would leave huge blocks of ice on walkways, prohibiting residents from leaving. This meant chopping up the ice as well as carting away the snow in wheelbarrows. Clearing walkways was critical. Quite a few of *kasetsu* residents were elderly, especially widows, those injured, and disabled. For these survivors, one slip on the ice or snow could break a knee, hip, or back and be fatal. Hospitals and nearby clinics were only now beginning to rebuild and regain their staff, personnel, and facilities.

Kasetsu were grouped in various sized clusters, housing anywhere from dozens to hundreds of survivors and their families. At one cluster of units in Otsuchi, we were preparing a community center, a place where occupants could free themselves from the isolation of their own *kasetsu* and get together for socializing and events. Here, older women would usually get together, drink tea, and chat. As scholar in psychosocial medicine Makiko Okuyama later shared with me in Tokyo, older women are more likely to become each other's support, getting together to talk, share their losses, histories, and feelings. Men, on the other hand, were less likely to do any of this. For this reason, the suicide rate post 3.11 eventually became higher among men.

One afternoon, at this same Otsuchi location, while some widows were serving us tea during our break from work, an old man who lived in a nearby *kasetsu* came to the community center and afterwards followed me around. He constantly asked how he could help me in my work. He needed to be physically useful. So, he and I worked together, lifting benches, chairs, cleaning windows, sweeping porches. Though I asked him a number of times for his name, he did not answer, though he asked for mine. From then on, whenever we would go to this site, he sought me out. He now at least felt useful, more alive. This was also his way of expressing gratitude to us for our work.

There were *kasetsu* spread throughout towns' edges—regular *kasetsu jūtaku* for survivors, *kasetsu* shops, *kasetsu* eateries or restaurants. Everything, everywhere was in makeshift mode. And with each day we gained a further taste of what life was like for these displaced survivors. We had glimpses of their fundamental disconnectedness from family, friends, and neighbors. They lived in a shadow world, situated literally in-between full recovery in their own home and utter loss. This literal in-betweenness is their Limbo. Our cleaning and sweeping the outside walkways to their *kasetsu* not only enabled survivors to walk and move outdoors more safely. It was also a way to reinforce their security. As described earlier, cleaning delineates a place that is secure from the threatening outside. And it helps in encouraging empowerment, redirecting control for those who have lost just about all control. A clean place is a safer place. As Brigitte Steger insightfully notes in her observations of shelter life, cleanliness = order = stability. Our shoveling was a shoveling away of Allison's "precarity." It helped in being rid of disorder, unsettledness, chaos, and external, unwanted interruption.

At Rikuzentakata, after a morning of working in drizzling rain, four of us had lunch together in one of the *kasetsu* shops—Ayaka Nishii, who studies English Literature (I call her "Aya." She and I would often have conversations about certain writers she loved), her friend, also Ayaka, who studies law, and Maki, who loves literature.[13] They talked about their lives and dreams upon finishing school. They all hope to come to America, not to live, but to visit its beautiful national parks and some cities like New York, New Orleans, and San Francisco, what they see in the movies. They promised to visit me. I asked each of them what they believed was the biggest challenge for survivors. Aya responded, "healing torn hearts." After we ate, we went to a nearby *kasetsu* candy shop just to browse before boarding our van to head back to the windy site at Rikuzentakata. Minutes after being in the shop, someone tapped me on the shoulder. It was the elderly man I spotted sitting in a chair when we came in. He offered me a cup of freshly brewed coffee that he purchased for me. I thanked him profusely, and he thanked all of us for doing our work. He then bought some boxes of candy and gave them to the ladies. We talked briefly as he shared with us how he had lost everything, his family and home, to the tsunami. He lives in a temporary shelter and regularly spends a part of his day in the company of others at the shop. He had little of his own. But his was a giant heart, and he gave what little he had.

Kagura

On another occasion while shoveling, neighbors were gathering together to witness a traditional festival dance, *kagura* (literally "god-performance"), for the new harvest of fish and eggs. It was held in what looked to be a private home now turned into a small community center. *Kagura* transformed it into a sacred space, inviting deities into that home to perform. Some people from nearby *kasetsu* also attended. That explained why the local TV network was there video-taping the event. A moment of heightened community interest, when survivors can gather together for a traditional ritual dance. We attended as well, taking a much-needed break from shoveling. The snow was particularly heavy that day, what we Americans call "heart-attack snow." We spent much time in back-breaking work chopping up steel-hard ice before we could get to the snow and cart it off.

This local celebration of *kagura* embodies in special ways the communal in-betweenness that brings together those who have suffered a shared plight. *Kagura* has its roots in Japan's ancient tradition of Shinto. Its dances are intended to celebrate our link to the rhythms of the seasons and our inescapable connection to and reliance on Nature's forces. Originally, these performances took place at sacred sites and were meant to invite deities (*kami*) to these sites. They also enable communities to gather together and renew their energies as a community, to pray for bountiful blessings for the village—a rich harvest, good health, good fortune, avoidance of disaster, and, should calamity occur, to relieve suffering. One thing for sure. They remind us of our indelible connection with the world of Nature and, for Japanese, *kami*. *Kagura* thus carry a profound spiritual meaning, a meaning that can be easily lost in more modern, urban Japan where these dances are often viewed as ancient oddities and tourist entertainment. This deeper spiritual meaning is also hard to preserve in a country with an aging population and thinning birthrate. But this is Tohoku, an area rich in tradition and one that cherishes and practices that tradition, despite its dwindling resources and fragmented communities. As Hiroko often reminds me, if I wish to experience the "real" Japan, then I can find it in Tohoku. My innkeeper friends, Toshihiro and Yuko Kuramoto, who run the Kuranoya Bed and Breakfast on the outskirts of Tono echoed that same sentiment. Even though he was quite successful working for Japan Newsweek in Tokyo, once Toshi and his wife visited Tono, they knew in their hearts that here they could tap into Japan's genuine spirit, away from the bustle of the city.

East Asian studies scholar Irit Averbuch says that in the Tohoku region, after 3.11, *kagura* now take on another meaning—a dance whose performance itself offers prayers for the dead, those swept away by the tsunami.[14] This ancient ritual dance, in reenergizing the local community, helps to heal its suffering. As a living prayer for a bountiful future harvest of rice, vegetables, and especially fish, it infuses renewed life into the people. *Kagura* depicts the survivors' in-betweenness, situated in-between their ancestral past with its honored traditions, and an uncertain future with numerous hurdles and hardships. Now, more so than ever, *kagura* is a time-honored ritual to help heal Tohoku's many broken communities. I was struck by the blend of generations witnessing the dance—children, young people, middle-aged and older adults. Here was a palpable social bonding that breathed life into all of us as we watched the dance with fascination and awe. We were united together in welcoming the deities into this space to give us renewed life and strengthen our bond as community.

As to these deities, it is not religious doctrine, teachings, or beliefs *per se* that tend to unite the community, but rather embodied expressiveness in festivals and dance. Are Japanese religious? This depends upon what we mean by "religious." If we think of religion as action that is church-going and dogma-oriented, the Japanese may seem less religious than we might expect. Many homes have private altars to remember deceased family members and pray to their spirits. These altars, *butsudan*, usually display photos, fruits, incense, and *omamori* (prayer momento for good fortune). They represent ways to bring hope to loss, order to chaos, renewal to ruin, hope to despair. All this means that the divine is not some distanced entity. In festivals, ritual, and dance, the gods are with the community. They dance among us. In this *kagura*, we watched as two demons danced, followed by a goddess with her beautiful fan dance, depicting the promise of good fortune and harvest. The three young lads I was working with, Takuya from Chiba and his classmates from Tokyo and Nara, were mesmerized, seeing *kagura* up close for the first time. Older women served us tea, oranges, and a delicious sake rice drink. Villagers allowed us, strangers, into their home, their space. I had this immeasurable sense of belonging, of being embraced unconditionally by these others who have lost family and friends. That evening the students were buzzing and sharing their excitement about this with others back at our base.

Against the Wind in Rikuzentakata

Working on Rikuzentakata's coast at that time of year, one word stands out—"wind," or *kaze*. The coastline there is known for its exceptionally strong winds. I grew up along the sea in "The Ocean State," Rhode Island, where I worked for years as a lifeguard at its beaches. Yet I cannot recall encountering the kind of unforgiving winds that nearly scattered us like toys and at times literally scorched us as we prepared the ground for the upcoming three-day 3.11 memorial ceremonies.

The wind was a major factor as well on 3.11. Winds carried radiation from the broken Fukushima Daiichi power plant to nearby mountain villages. And, not far south of Rikuzentakata, as waves and fire swept through the small island of Oshima, winds caused the fire to surround the island, making it nearly impossible for rescue boats to deliver food and supplies. Aid finally came to the island inhabitants due to the collaborative efforts of the SDF and the U.S. Marines' Operation Tomodachi, "Operation Friendship." Through Operation Tomodachi, American forces stationed in Japan provided invaluable assistance in what has been dubbed "the single largest humanitarian relief effort in American history."[15] American Marines worked together with the SDF in helping the island's inhabitants, who numbered over 3,000.

Preparing for the ceremonies at Rikuzentakata meant digging into the rock-hard ground, loosening the soil enough to put in lanterns, stakes for lights, stringing up the lights amidst gale winds, pounding metal posts into the ground for wooden platforms and frames for signs, further lights, and candles. The wind screamed at us, whipping us around like scarecrows, while we doggedly bent forward chopping, digging, hauling, dumping, chopping, digging, hauling, dumping. All this time, with the wind stinging our faces, we struggled to make headway, one shovelful at a time. We labored together and held our ground.

The Shoe

In the course of digging into icy ground, I came across the top half of a shoe frozen in the soil. It was clearly a man's shoe, size 9 or 10, and immovable. A few of us struggled in vain to pull the rest of the shoe out of the ground. The cold, hard ground claimed ownership. The earth always wins. On some of my many motorcycle jaunts while studying in Belgium, I would visit my Irish cousins in Swinford, County Mayo, home to my grandfather and his sister Bessie who had lived in a small cottage outside of town. I'll always remember a headstone in the tiny town's cemetery overlooking the village that read, "God's green earth is a warm bed." The man's shoe had a mind of its own. Or the owner may have been gripping it from the other end. In any case, it stubbornly refused to come out of its resting place.

I have little doubt that the shoe is still there. So, amidst our stakes and lights for the ceremony, there lies the immutable shoe. It seems intended to remind us of the Buddhist *dukkha*, often translated as "suffering" but which literally means "dislocated." Suffering is in essence a dislocation from our genuine condition. Dislocation comes in various forms. And the shoe stubbornly reminds us of Nature's crushing blow and, at the same time, a human itch to stoically ride out the storm, whispering silently in utter defiance "I'm still here. You can't take all of me." Something about these objects. Objects carry an intimate relationship with memory. Naturally, we often think of memory as referring to the past. Yet memory carves meaning in the present. The term "remembrance" denotes a cracking into the surface of the present of that which occurred in the past, and so this memory keeps the past alive. And through objects associated with the deceased, memory has a relational quality not only to the past but to both present and future. Japanese anthropologist Kazuhiko Komatsu equates the departed's "spirit" with the memory of the departed. In a sense they are one and the same. Spirit is memory. Memory keeps alive the spirit.[16]

As the Pacific War (1941–1945) neared its end, funeral urns supposedly containing the remains of soldiers were returned to families. However, it was discovered that most of these urns housed stones or pieces of wood instead of ashes or bones. Japanese military officials explained that most bodies simply could not be retrieved. Nonetheless, the spirits of the dead soldiers were still somehow present in the urns. And by praying to the spirit of the deceased, the bereaved family members kept their memory alive and comforted them on their after-death journey. An example of this is the fascinating account written by an officer in Japan's 29th Foot Regiment, "The Memorandum of Sand of the Fallen Soldiers' Spirit." Soon after the bloody struggle of Guadalcanal in 1942, a turning point in the Pacific War with countless casualties on both sides, it was impossible to send back remains of the soldiers. Instead, after holding funeral rites at various spots where soldiers died, survivors gathered up coral sand from the beach, and equally distributing grains of sand according to the numbers killed, placed sand in each urn to be sent to bereaved families. Families expected remains in some form in these urns, like ashes, bones, or hair. Military officials, however, rationalized doing this by asserting to families that the sand embodied the spirits of the dead.[17] They were sending back to families the spirit of their deceased.

Objects take on a special meaning when their owners are violently pulled away from the land of the living, yanked from existence. People have lived with these objects, worn them, clasped them, washed them, and handled them daily.[18] Throughout our work, we occasionally came across objects imbedded in the ground. A glove, hat, clothing, a broken tea cup, a splintered ceramic plate. I retrieved a piece of porcelain buried partially in a barren field in Rikuzentakata. It resembles part of a small dish or bowl. This is my own *omoide no mono*, "memory thing," a way of keeping the past alive in the heart. This piece of someone's life is now accidentally bequeathed to me. It stays in my office at home, a reminder of how fragile life is. We can be so easily torn away. But our hearts will remain unbroken. My porcelain piece reminds me of what really matters in life. And what we love is never lost.

The Miracle Tree

As we described earlier, Rikuzentakata is where the "miracle tree" stands as the sole pine tree that survived the waves. It is a symbol of hope for survivors. Within months after the disaster, it died, its roots finally giving way. A prosthesis has now replaced the dead portions of the tree. Though artificial, the symbol still matters. Maintaining this symbol, however, comes at high cost. It was enormously expensive to raise the ground to keep the tall tree firmly planted. Locals remain incensed at the costs. They are skeptical about allotting scarce fiscal resources to this monument when the monies could be used in better ways, particularly for those in *kasetsu* eking out each day and those local shopkeepers who struggle to keep their businesses alive. This is similar to concerns Otsuchi citizen have towards proposals to build a memorial monument to the victims of the disaster, which will be quite costly. They argue that funds need to be spent for people, jobs, and improved shelters. Critics also claim that erecting these symbols is politically driven, a way for local governments to convey an image of a Japan that is now in recovery mode, "back to normal."

At Rikuzentakata there also stands a gutted building close to shore which remarkably withstood the waves. Local officials have asked the owner that it be torn down. To residents, it's an eyesore and painful reminder. The owner insists, however, that it remain standing as a symbol of resilience and hope. Whether the officials or the owner will win out, it is still standing, and the owner pays an exorbitant sum to the government to keep it upright. The hollowed out building still stands there, reminiscent of the famed Hiroshima Peace Memorial, or Genbaku (Atomic bomb) Dome. The Dome is the only structure left standing near the hypocenter where the B-29 *Enola Gay* dropped the bomb on Hiroshima on August 6, 1945. All-out efforts by citizens and officials have kept the structure in its original condition. It serves as a twofold symbol: first, of the terrible and destructive power of nuclear weapons, and second, as a sign of hope for the elimination of such weapons and the cultivation of world peace.

Hakozaki, 11 March 2014

The seven of us made our way through curving mountain roads, occasionally catching peaks of the stunning coastline with its blue sea against densely treed steep hills. This was the morning of March 11, the day the disaster struck three years ago. As we climbed down towards the road's end, we spotted what remained of the small town of Hakozaki, just outside of Kamaishi City. One would never guess that the waves would hit the town with such hammering force, crushing two-thirds of its homes. From all appearances, the town seemed safely snuggled, harbored in a bay with an inlet well-shielded by fingers of hilly land to act as natural barriers. However, giant, formidable waves from the open sea arrived with such vengeance that they literally swallowed all ships in the harbor and leapt easily over the thirty-foot seawall protecting the eastern face of the town, crushing homes and inhabitants. Those on higher ground were saved from the hammer. Figure 6.3 shows the leveling of more of Hakozaki except for buildings on higher ground.

Upon reaching the town at ground level, we walked through what was once a quaint, little town and later to the water's edge of the harbor. At exactly 2:46 p.m., the time the earthquake struck, the sirens sounded for five minutes. We stood in reverence. We then each placed a yellow carnation into

Figure 6.3. Hakozaki, 11 March 2014

the water and quietly watched as all seven flowers gently floated away. We prayed in silence, bowing with our hands together. Sudo, our group leader that day, hails from Hokkaido and regularly helps lead the volunteers at this time of year. He likes his sake and beer. And he took me by surprise when he made what looked like the sign of the cross. I wonder if he did this out of respect for my Christian background. As we walked back to the van, Sudo pointed to the tunnel leading into the town. There, over three years ago, hundreds of corpses were trapped inside until the waters receded. Of the 240 homes in Hakozaki, 160 were washed away. Throughout that day, we felt a thundering silence.

We Are What We Give of Ourselves

All we own is what we give away. On one particularly grueling day shoveling snow in Kamaishi, our leader Tetsuya introduced us to the Kobayashis, an elderly couple in their mid-eighties. Their home, a small traditional Japanese style wooden structure, rested on a hilly road just on the outskirts of the city. Though a few miles away from the coast, they barely managed to escape the giant waves that swamped Kamaishi. The two had lost family, friends, and neighbors to the tsunami. They were among the lucky few to escape the waves, which literally licked their doorstep. After meeting us at the door, they invited us into their home for tea and snacks. There we were, nine of us, crammed into their tiny living room. Yet, it was exceptionally comfortable, wrapped in the warmth of steady heat from its small kerosene heater. This was the welcome heat that we all so missed. Tiny Mrs. Kobayashi was bustling over with energy. Without stop, she served us tea, rice crackers, snacks, and coffee. Her husband sat in his armchair with a giant smile and talked about the rugby match on his small television. Japanese love rugby, and Kamaishi has its own professional rugby team, ironically named the Kamaishi Seawaves RFC (Rugby Football Club). He and his wife profusely showered us with their generosity. The warm room was a much needed respite and shelter from the incessant and brutal cold outside. The Kobayashis did not have much. But what they did have, they shared with us, strangers. And when we left to return to work, they stepped out into the cold to bid us a warm farewell. The husband managed to totter carefully down the road to see us off, waving goodbye.

Ayumi Kakuhari is in her mid-twenties. She lost her younger sister to the tsunami. She watched from a distance as the giant wave swallowed her home, street, and neighbors. When we arrived at Rikuzentakata to clean up after 3.11's three-day ceremony, she led the local volunteer group also there to help. After leading us through *rajio taiso*, Ayumi rushed up to me and embraced me in a warm, welcome, heartfelt hug, quite uncharacteristic for Japanese. Her round face lit up with a gracious smile. Gentle displays of deep affection and gratitude usually occur through grasping another's hand with both hands, as Usuzawa did with me after his talk to volunteers in Tono. Ayumi's dark brown and sad eyes looked straight into me. They reflected a warm, kind heart. This was the kind heart that spoke to me throughout my stay in Tohoku, the heart evidenced in volunteers with whom I worked, a heart strong and innocent, a heart that warmed our bitter cold and windy days and nights. We then talked and she told me about her sister. She later sent me a brief note that she delivered to our leader Keita, with words that contained deep gratitude as well as a cry for help to work together to build a future.

We are measured by what we give of ourselves. Mikiko Sugawara personifies the remarkable in the ordinary. She is a humble, hard-working mother fully committed to caring for her small family. Her spirit is incredibly strong. Charismatic, her inner strength adds to her physical attractiveness. She managed to serve tea to fifteen of us in her small shelter. I was struck by the grace with which she boiled the water and passed out the cups. She never smiled, harboring a deep, hidden loss. Most of her neighbors were swept away by the waters, and she had only a few family left to care for. Yet her actions smiled for her, and exuded her warm heart. Mikiko took on the task of coordinating the care of numerous survivors, over nine hundred, in Rikuzentakata, many of them elderly and disabled who were placed in the few remaining shelters after the tsunami. She remains a truly exceptional person and leader, and embodies a strong *elan vital*, a life force that prevails among people in coastal communities still struggling to regain footing.

Not all volunteers with whom I worked were university students. Katsunori Kumagaya (*kumagaya* literally means "strong mountain"), looked to be in his mid-forties. He helps supervise the Ichihara City Sports Association in Chiba Prefecture. Katsunori was a "repeater," having worked with Magokoro Net each year since the tragedy. As with all of us, it was his small way of helping the survivors and building a better future. An athlete with a strong interest in rugby and soccer, he was a former physical therapy trainer. French philosopher Albert Camus, another ardent soccer player, once claimed that everything he learned about ethics came from playing sports. Katsunori and I shoveled snow together at a good pace. He taught me about the social significance of sports in some of the coastal communities like Kamaishi, a way of holding communities, particularly broken communities, together. Kamaishi, with its Seawaves, is vying as the site for the 2019 Rugby World Cup. But it is competing with Osaka. Given Kamaishi's fractured infrastructure, Osaka will most likely win the bid. Nonetheless, in the aftermath of disaster, sports is assuredly a source of community pride and recovery. Throughout the *kasetsu* shops were pins, t-shirts, photos, and memorabilia of the Kamaishi Seawaves.

Yosaku Oshiro, or "Saku," twenty-eight years old, hails from Okinawa and is another repeater, having come twice a year since the disaster. He shared with me his enduring need to help the marginalized. For Saku, volunteering promises to become his life work. He now works for the Association for Aid and Relief, Japan (AAR Japan), an International Non-Government Organization (INGO). He is currently in Yangon, Myanmar, managing a vocational training center for persons with disabilities. Many of those physically handicapped are polio and landmine victims. Myanmar has one of the world's highest numbers of landmines. For Saku, his chosen path is no detour from the routine.

It is often the case that our connectedness only becomes apparent just before departures, like plane passengers seated next to each other who finally speak with each other when the plane touches ground. Imminent leave-taking is permission to make a brief and only encounter. Yet there was something about Sayoko Kawabata. She seemed to have an intimate grasp of how things unfolded at Magokoro Net. I spoke with her just hours before she and her small group of students left for Tokyo. Sayoko teaches occupational therapy for the faculty of Health Science Technology at Bunkyo Gakuin University in Saitama, northwest of Tokyo. She has lived in the United States for some years, and speaks excellent English. Sayoko shared with me the events of the 1995 Kobe earthquake and how that changed her life. Since 3.11, she regularly brings a group of her students to Magokoro Net to volunteer.[19] She tells me she will bring her students again in 2015 and will be there for the 3.11 memorial ceremonies.

Give us this day our daily grind. All this—clogging together though the fickle wind, cold, and wet, shoveling, cutting wood, removing rocks, hacking through ice—embody precious lessons in what I think of as the "virtue of presence." This is the simple truth that amidst chaos, despair, and anguish, we can communicate in simple, straightforward, yet eloquently profound ways, not through talk, but by genuinely being there, being with, and being for. With each day's journey to the *kasetsu*, we the living worked alongside survivors in solidarity, sharing in their collective pain. Working together amidst the ruins, we became more alive.

It is quite profound, this day-to-day working as a group. It reaffirmed that we are in this together, and together we will rebuild (*fukkō*) and recover (*fukkyū*). Hence the meaning behind the ubiquitous "V" sign whenever we took photos of each other—*Nippon Ganbarō*—"Courage Japan." We will endure. Together we survive; together we thrive. In a way, we volunteers are also offering our own *kuyō*, or service for the departed; our own drawn-out ceremony to help soothe the narrow path each soul travels *en route* to the other side. I felt especially closer to my roots. Here, I came in touch with my ancestors, and Mom and Dad were there with me, smiling, and urging me on. With every *kampai*, it is to those who leave the world to us, present and future.

By any measure, our work is ordinary but not mundane. Consider our shoveling snow off roads, walkways, steps, and other areas around the numerous *kasetsu jūtaku*. A snowplow, if it could manage access to these areas, would certainly be more efficient, quicker, and back-saving. But this cannot replace the human-to-human connection of working together with our shovels. It was not the work as such. It was our collective presence working alongside with and for the suffering survivors. For many of us, in the presence of the dead victims' spirits, that makes the difference. Our shoveling is a shoveling away indifference and irrelevance. The paths we clear for walking are paths that link together those within and those outside these clusters of *kasetsu*. We shovel so that survivors can connect with others and keep on living. In his poem "On Friendship," Kahlil Gibran writes, " . . . for in the dew of little things the heart finds its morning and is refreshed."[20]

NOTES

1. I offer a kind thanks to Magokoro Net's public affairs manager, Makoto Yanagisawa, for furnishing details.

2. Peggy Fan, "Building a Resilient Volunteer Organization," *Stanford Social Innovation Review* (March 10, 2014), accessed Mar. 3, 2015, http://www.ssireview.org/blog/entry/building_a_resilient_volunteer_organization.

3. Peggy Fan, "Building a Resilient Volunteer Organization."

4. Tom Wilson, "In Japan's Disaster Zone, Recovery Is Just Beginning," *Time* (Oct. 26, 2011), accessed March 3, 2015, http://content.time.com/time/world/article/0,8599,2097511,00.html.

5. Expanded from Michael C. Brannigan (Kenji Kimura), "One Shovelful at a Time—Volunteering in Tohoku," *Japan Railway & Transport Review* 65 (March 2015): 44–45.

6. Lucy Birmingham and David McNeil, *Strong in the Rain: Surviving Japan's Earthquake, Tsunami, and Fukushima Nuclear Disaster* (New York: Palgrave Macmillan, 2012), 136.

7. Birmingham and McNeil, *Strong in the Rain*, 137.

8. Video interview with Ryoichi Usuzawa, by Tono Magokoro Net Film Team, "Great East Japan Earthquake Series, Witnesses of the Tsunami, Ryoichi Usuzawa" (Tono Magokoro Net, 2012), accessed March 5, 2015, http://translate.google.com/translate?hl=en&sl=ja&u=http://tonomagokoro.net/&prev=search.

9. Tono Magokoro Net Film Team, "Great East Japan Earthquake Series," http://translate.google.com/translate?hl=en&sl=ja&u=http://tonomagokoro.net/&prev=search.

10. Stanley Cohen, *States of Denial: Knowing about Atrocities and Suffering* (Cambridge, UK: Polity Press, Blackwell Publishing, 2001), 259.

11. Leslie M. Tkach-Kawasaki, "March 11, 2011 Online: Comparing Japanese newspaper websites and international news websites," in Jeff Kingston, ed., *Natural Disaster and Nuclear Crisis in Japan: Response and Recovery after Japan's 3/11* (London: Routledge, 2012), 121.

12. Brigitte Steger, "Solidarity and Distinction through Practices of Cleanliness in Tsunami Evacuation Shelters in Yamada, Iwate Prefecture," in *Japan Copes with Calamity: Ethnographies of the Earthquake, Tsunami and Nuclear Disasters of March 2011*, 2nd ed., edited by Tom Gill, Brigitte Steger, and David H. Slater (Bern, Switzerland: Peter Lang, 2013, 2014), 53–54. Professor Steger is currently working on a book describing "shelter life" for survivors.

13. All this time I was intensely aware of my back when shoveling, chopping, and lifting. Aya knew about my back and, while we were all working, would at times check up on me to make sure I was alright from all the shoveling.

14. Irit Averbuch, "Modern Meanings of Kagura Performances," *Dharma World* (April-June 2013), accessed Dec. 11, 2014, http://www.rk-world.org/dharmaworld/dw_2013aprjune_modern-meanings-of-kagura-performances.aspx. Also accessed Dec. 11, 2014, http://www.xip.fi/atd/japan/kagura-sacred-shinto-dances.html.

15. Birmingham and McNeil, *Strong in the Rain*, 133.

16. Akira Nishimura, "Battlefield Pilgrimage and Performative Memory: Contained Souls of Soldiers in Sites, Ashes, and Buddha Statues," *Memory Connection* 1, no. 1; The Memory Waka, (2011): 304, accessed Jan. 15, 2015, http://memoryconnection.org/wp-content/uploads/2011/12/AkiraNishimura1.pdf.

17. Nishimura, "Battlefield Pilgrimage," 304.

18. For a fascinating account of the significance of common objects in everyday life, see Sherry Turkle, ed., *Evocative Objects* (Cambridge, MA: MIT Press, 2007). See also Anne Allison, *Precarious Japan* (Durham, NC: Duke University Press, 2013) 195ff.

19. Upon return to her university after volunteering, Sayoko put me in contact with her colleague, Professor Hiroko Shimazaki, who does close research on the plight of survivors living in *kasetsu*.

20. Kahlil Gibran, *The Prophet* (New York: Alfred A. Knopf, Inc., 1923), 59.

Chapter Seven

Fukushima's Unseen Monster

Watching the tragedy unfold at the Fukushima Daiichi Nuclear Power Plant on his television, seventy-two-year-old Yamada Yasuteru was horrified to see young workers in their hazmat suits risking death in order to stabilize the damaged reactors. The plant workers were clearly in between a rock and a hard place. They risked dying from an explosion and from the deadly radiation that was smoldering within the facility. They had no choice but to open vents by hand in order to release deadly steam into the atmosphere. Otherwise, an explosion would have incurred even more damage from radiation. Yamada, a retired Sumitomo Metal Industries plant engineer, knew that he had to take action.

WE ARE NOT KAMIKAZE

Without delay, Yamada leapt from being a retired observer of the events of March 11 to actively lead his peers in a bold cause—to help stabilize the damaged nuclear plant and to act in place of the younger workers, regardless of risks to themselves. His reasoning made perfect logical sense. They had the experience and requisite skills. They were also at less risk of diseases that would come from overexposure. Due to their age, they were less vulnerable to contamination. Even if contamination led to cancer and other diseases, this would likely take years to develop. Having himself survived lymphoma, Yamada states, "I want to make the most of the time I have left." In all probability, his group would die well before cancer takes its grip. For Yamada, all this pointed in one obvious direction: they, older and retired, should step in to help relieve the younger workers; engineers, technicians, and cleaning-crews still have their lives to live, with responsibilities for their families.

Yamada took full advantage of Twitter and email in contacting friends, colleagues, and former coworkers. He even formed his own blog through which he made persistent appeals to his generation.[1] Through this power of connectivity, he was able to reach out and seek fellow support among retired engineers like himself, workers at power stations, and factory design experts. It will be their person-to-person connectedness that will enable them to face up to the challenge together.

The calamity at the nuclear plant continued to unfold. In April, with his longtime friend Nobuhiro Shiotani, another retired scientist, Yamada co-founded the Skilled Veterans Corps for Fukushima (SVCF), retired engineers all over the age of sixty. They then lobbied persistently to the government with a measured and detailed plan for the group to work in the damaged plant. Though they were at first refused, they persisted in their offer to help. They aimed to help mitigate the damage at the Daiichi Nuclear Power Plant by taking an active part in contamination clean-up at the impaired nuclear reactors. Its official proposal requests:

> Substituting unnecessary radiation exposure of the young labor engaged in the cleanup works of the damaged Plant by the introduction of elderly persons to the best of their ability in these tasks.[2]

This overall aim includes monitoring levels of contamination and helping with removal of debris in contaminated areas. Moreover, the SVCF, which currently has over two thousand members, plans follow-up measures to remove reactors that are damaged and help see to it that contaminated waste is safely deposited. The group is particularly keen on not leaving these tasks to the management of privately owned TEPCO. The SVCF consists of retired men *and* women, like Kazuko Sasaki, a seventy-two-year-old grandmother. For Kazuko and many others, motives include a sense of collective responsibility and duty. She points this out clearly: "My generation built these nuclear plants. So we have to take responsibility for them. We can't dump this on the next generation."[3]

Kazuko's sentiment and the work of the SVCF illustrates Watsuji's emphasis on our indelible in-between. And this is very much in line with Emmanuel Levinas's emphasis on living not simply toward our own individual deaths, but towards a life beyond our own deaths. For Levinas, one major flaw in Martin Heidegger's (1889–1976) ontology is that I am essentially, to paraphrase Heidegger, a "being toward *my own* death" (author emphasis). For Levinas, we need to think of ourselves as beings toward a life on this earth beyond our own deaths. He writes:

> Heidegger deduces all conceivable meaning from the attitude of man toward his own death. He thinks to the very end, in the senses of the term. He carries out his thought to its ultimate consequences, and he thinks that my death for

> me can be nothing but the ultimate self. . . . Is there not a manner of thinking that goes beyond my own death to the death of the other man, and does not the human consist precisely in this thinking beyond one's own death?[4]

Recognizing and honoring our in-between, our *aida*, naturally entails this. Our in-betweenness is temporal as well as spatial. Situated between past and future, our present—which always is—unfolds into the future, which is about to be present. The increase in so-called "acts of God" demands that we look well beyond our own individual deaths.

SVCF members have freely joined, without coercion or strong pressure. Their solidarity consists in their free choice, regardless of their own personal, political, or ideological motives. They remain convinced that their "objective aimed at the completion of cleanup works in the stricken nuclear power plant highly magnetizes and unites a wealth of members."[5] The government appreciated their willingness to work at the plant. Yet officials remain cautious about letting them into the contaminated area. Though they eventually declined their offer, Yamada and his group represent a remarkably concrete and poignant commitment that acknowledges our inescapable in-between. "It is our fundamental principle to challenge an unprecedented disaster by means of our amassed profound knowledge and experienced skills, while we unite as one, in mutual respect."[6]

Yamada, who was then honorary president of SVCF, died on June 17, 2014. He once said that "Most Japanese have this feeling in their heart. The question is whether you step forward, or you stay behind and watch."[7] Here we see *aida* beautifully at play. For the young workers, their in-between is flanked by their youth and young adulthood and their unfolding middle age. For Yamada and his generation, they are in-between a full life already lived, their elderly years and their deaths. And, in the long run, what really matters is what will benefit their society now and in the future. At the same time, he is realistic and cautious in his approach. By no means is he vain, seeking attention, and eager to die. As he once said, "We are not kamikaze . . . They were going to die, but we are going to come back."[8] It is not desire for personal glory that drives his group's proposal, but their strong sense of duty and selfless responsibility to do what they can for the common good.

AN UNFOLDING NIGHTMARE

Here is one of history's darkest ironies. Japanese remain the only targeted victims of nuclear attacks—Hiroshima and Nagasaki. Long-range effects on survivors and their families were devastating. Since then, Japan has constructed an energy policy that has led to full-fledged commitment to nuclear power and dependency. Now, its worst nuclear accident from its 3.11 triple disaster will generate for numerous survivors far-reaching adverse consequences for decades to come as they face an uncertain future. The name "Fukushima" combines the terms *fuku*, or "good fortune," and *shima*, "island," meaning "island of good fortune." Instead, Fukushima's calamity is a terrifying unraveling of a natural disaster in a socio-technical environment combined with a tragic litany of errors, oversights, mismanagement, and outright deception that will forever scar the country and its people. It also raises the crucial question whether the crisis will alter Japan's future energy policy.

Here's a skeleton of what happened. As the world knows, 3.11's colossal earthquake spawned a titanic tsunami that easily breached the protective seawall at the Fukushima Daiichi Reactor complex. Clearly, the plant did not learn from the 1707 Hoei and 1933 earthquakes that produced massive tsunamis with waves well over ninety feet high. Nor did it learn from its recent past. Between 2002 and 2006, the nuclear power plant had a blemished history of all sorts of technical faults, resulting in over twenty reported problems.[9] Within minutes, waves flooded generators and back-up cooling systems.

Experts now claim that the earthquake itself was so powerful that it impaired the cooling system in reactor one even before the onrush of the tsunami. Once fuel rods melted down, radiation contaminated groundwater, soil, and sea. Scorching temperatures produced a series of explosions that released radiation and toxic gases into the atmosphere. Within minutes, the unimaginable was becoming a reality. A hellish nightmare unfolded: total power failure, loss of back-up generators, release of deadly radiation, nuclear meltdown, and with it the danger of nuclear fission and meltdown at nearby plants and beyond. Not just Japan, but the world was at risk.

With each day, matters were getting out of control. The loss of all power was a dominant factor. Workers scrambled around parking lots and elsewhere searching for functional car batteries. They could only manually operate key vents in only so much time before dangerous radiation exposure. On March 12, the first hydrogen explosion takes place in the unit 1 reactor building. Radiation spewed into the atmosphere, and winds carried radiation north and northwest over Minamisoma, its farm areas and forests, settling especially in the small mountain villages of Namie and Iitate. (Iitate, now a nuclear ghost town, is the village we drove through in 2014 to meet with Mayor Sakurai in Minamisoma.) On March 14, a second hydrogen blast ripped open the building protecting reactor three.

Early in the morning on March 15, upon hearing from his economic minister Banri Kaieda that the Daiichi workers were seriously considering abandoning the severely damaged plant, Prime Minister Naoto Kan realized the apocalyptic consequences of having the abandoned plant spin out of control, probably sending poisonous air to Tokyo and incurring mind-boggling fatalities among the thirty-five million Tokyoites.[10] The prime minister immediately flew by helicopter to a protective bunker at the nuclear site which acted as the operational base for the stricken plant. He implored workers to stay, appealing to their sense of duty, *giri*, to their country and to the world. In a sense, he was reminding them of their inescapable in-between. The situatedness of the plant meant that not only was the plant's safety at stake, along with Fukushima Prefecture and all of Japan. What was at stake, implored Kan, was the safety of the world. He then immediately ordered plant engineers to hold their ground, stay in the facility, and open the vents, which meant doing it manually as systems were all shut down. They could only work in shifts of seventeen minutes. Beyond that, they were exposing themselves to fatal radiation levels. His appeal was the customary call to duty, *Onegaishimasu: ganbatte kudasai.* No doubt, Kan's aggressive actions, bold decision, and his point-blank order incurred hostility within Tokyo Electric Power Company (TEPCO), the owner and manager of the Daiichi power plant and government ranks. Moreover, whether or not the plant was to be entirely abandoned remains a question mark. Yet deliberating on the appropriateness of his actions at that time is somewhat moot. A sort of Monday-morning quarterbacking after the fact. Given the information he had at his disposal, he had to act quickly.

So did workers attending to the fuel pools. With the sea overcoming the coastlines and ravaging its towns, water was everywhere. But in the reactors' fuel pools, water was in short supply. Japan did not face their plight alone. For instance, the United States offered invaluable assistance. During the early stages of the fallout of the Daiichi power plant, U.S. unmanned drones flew over the plant to photograph and collect data to send to the Japanese government. The U.S. Navy also provided nearly two million liters of water to help cool the overheating reactors.[11] All this clearly demonstrates our natural in-betweenness, our interconnectedness. Despite the ongoing dissension regarding the continued presence of American bases in Japan, especially in Okinawa, the positive memory of U.S. Marines no doubt paved a path for more favorable perceptions of Americans in Japan. All is cause and effect.

There were endless efforts, 24/7, to save the dying nuclear power plant. For instance, consider Tokyo's elite force of firefighters. They and others represent vignettes of moral grit—courage, faith, and duty. The government ordered a select group of firefighers to enter the Daiichi plant as first responders and to immediately cool down the water-starved fuel pools and their overheating fuel rods. These were only men over forty and who already had children. Temperatures of the fuel pools were rising at alarming speed. The pools needed water immediately. The firefighters faced the daunting challenge of connecting series of long and heavy hoses by hand to the reactors from the sea, over a quarter mile away. This meant stumbling and working across a radiation-filled killing field of black water, mud, bodies, debris, sharp metal and glass. These elite painfully and slowly succeeded, paving the way for back-up crews, plant and recruited workers to continue laying miles of pipes from sea to reactor. Lethal radiation levels demanded that they work in hazard material suits, masks, and gloves. In April, the government raised the radiation scale, measured by the International Nuclear and Radiological Event Scale, to level 7, the same level as the 1986 Chernobyl crisis.[12] The firefighters continued their work into the Spring and Summer, when temperatures were blisteringly hot, especially with HAZMAT gear.

Mainstream Media Toned Down

Government officials, TEPCO, and media joined hands in toning down the gravity of the events at the nuclear complex. Government spokesman chief cabinet secretary Yukio Edano, who became a familiar face throughout the unfolding tragedy, continued to play down the nuclear facility's damage and lethal threat to the region. Both Edano and TEPCO were complicit in toning down the facts. With Edano's public pronouncement on March 16 that "Radiation levels around the nuclear plant are not at levels to cause an immediate health risk," the vagueness of the comment along with inconsistent remarks from radiation experts heightened suspicions among the public. Moreover, the "M" word, "meltdown," was anathema in industry and government conversations. Indeed, there was an all-out effort on the part of government, industry, and mainstream media to avoid using the term. This extra caution was no doubt a reaction to excessive hyperbole in foreign media accounts, many of which were alarmist like the UK's popular tabloid *Sun* describing Tokyo as a "city of ghosts." Foreign accounts were often sensationalized and distorted. Temple University's Jeff Kingston, who heads the Asian studies program on the Japan campus, pointedly criticized much foreign reporting, filled with hyperbole and instances of "parachute journalism."[13]

There is a cautionary subtext in all this. Despite the exaggerated accounts, it became clear that mainstream media was pretty much in the pocket of the official voices, the government and the nuclear industry. As Tokyo-based correspondent David McNeill writes in his co-authored *Strong in the Rain*, the Nikkei Advertising Research Institute reported that prior to the disaster, the nuclear power industry was "Japan's biggest advertiser, spending 88 billion yen (roughly US $1 billion) a year."[14] Mainstream media therefore reported the facts as they were filtered down to journalists. Their reporting was descriptive and hardly investigative. Not atypical, as there is little in the way of investigative reporting in Japan. A striking example of this is the powerfully influential public service broadcasting station NHK. NHK is a huge media force throughout Japan and remains globally well-known and respected for its descriptive reporting. However, throughout the crisis, the information it provided was funneled through government and industry sources, often offering commentaries from pronuclear experts but providing little in the way of voices from nuclear critics. NHK's descriptive reporting, though helpful, was not balanced with information from other angles and sound analysis.[15] People were therefore not sufficiently informed as to the unfolding of the crisis.

The crisis also revealed the unique workforce culture among mainstream media companies which viewed their reporters as first and foremost employees under their hire *and protection*. Reporters were therefore prohibited from entering dangerous areas, not because of what facts they may unearth, but because they were part of the company, and thus under the protective cloak of the company. For their own safety, officials did not allow them to enter the twelve-mile exclusion zone around the plant. Moreover, by mutually agreeing amongst themselves to act this way, there was little likelihood that any one of the major media companies would trump the other in scooping a story. All this results in soft-pedaling the facts. I wrote about this in my April 26, 2014, *Albany Times Union* ethics column, "Just how safe is safe?"[16] This downplaying led to the deliberate omission of critical facts. Japan's Nuclear and Industrial Safety Agency (NISA) withheld facts from the general public about the direction of the radioactive plume from the March 12 hydrogen explosion on the grounds of preventing widespread panic. Without such information, however, thousands who fled the nuclear plant epicenter ventured into towns that were clearly contaminated. This "downplaying" of the facts, akin to deliberate deception, is nothing new in the wake of nuclear disasters. The same occurred with Chernobyl, to this day the world's worst nuclear accident, and Pennsylvania's Three Mile Island. First, a quick glimpse at Three Mile Island.

On March 28, 1979, a perilous recipe of equipment glitches, worker slip-ups, and design problems generated the most serious U.S. commercial nuclear power plant operating accident. The Three Mile Island (TMI) Unit 2 reactor, close to Middletown, PA, partially melted down. "Partially" is the operative word here. It was later ascertained that nearly half of the nuclear core melted. Soon after the accident official estimates of the damage were a litany of deception, downplaying the reality. Officials claimed first the absence of radiation release, then "control" of what was released, then indicating that the release was "insignificant," and other untruths such as no real risk of explosion and no pressing need to evacuate nearby communities. It took nine years before the plant's owners admitted there was a partial meltdown, even after there was evidence of mounting rates of cancer, birth defects, and leukemia among local residents living within a ten-mile radius of the plant.[17] In his insightful analysis, environmental journalist Gar Smith writes:

> It wasn't until 1983 that a sonar probe revealed extensive damage caused by a partial meltdown that destroyed nearly half the reactor core. TMI's containment vessel had failed, releasing a cloud of radiation that was about 100 times more significant than the initial estimates offered by the industry and the NRC [Nuclear Regulatory Commission].[18]

All this later led to tighter steps to enhance regulatory measures and reactor safety. Yet, despite strong public opposition to the plant's continued operation from surrounding communities, including the state capital, Harrisburg, the plant is still active.

As for Chernobyl, when a dramatic power surge led to an explosion of the Chernobyl nuclear power complex's reactor Unit 4, massive amounts of radiation, especially cesium 137, was released. Estimates as to the amount of radiation released report around 50 percent of the iodine and cesium contained in the reactor core. According to the World Nuclear Association,

> The accident caused the largest uncontrolled radioactive release into the environment ever recorded for any civilian operation, and large amounts of radioactive substances were released into the air for about 10 days.[19]

Two workers died from the explosion, and twenty-eight eventually died from acute radiation syndrome (ARS). Hundreds of other workers as well as millions of residents in the area, including parts of Ukraine, Belarus, and the Russian Federation, were exposed to high levels of radiation. This generated the evacuation of nearby Pripyat and those living in the thirty kilometers (18.6 miles) exclusion zone and the resettling of an additional 200,000, in total, over 320,000 people. Close to 6,000 children, many of whom drank milk infected with radioactive iodine, were later detected with thyroid cancer. Fortunately, almost all survived following treatment.[20] Critics claim that the number of fatalities have been played down. Instead of the official number of thirty-one "liquidators" (cleanup crew) killed, recent figures tell a different story.

By 2005, between 112,000 and 125,000 of these "liquidators" have succumbed to radiation exposure. Over the past twenty-three years, according to the most recent estimates, the Chernobyl blast may have caused the cancer deaths of nearly a million people worldwide.[21]

In August, 1986, the International Atomic Energy Agency held a special meeting in Vienna, Austria, to review the shocking Chernobyl nuclear disaster months earlier on April 26. Renowned scientist, Valery Legasov, led the Soviet delegation. Up until then, blame for the accident was targeted towards various parties, especially towards the plant's operators. Legasov offered his own honest assessment:

> After I had visited Chernobyl NPP [Nuclear Power Plant] I came to the conclusion that the accident was the inevitable apotheosis of the economic system which had been developed in the USSR over many decades. Neglect by the scientific management and the designers was everywhere with no attention being paid to the condition of instruments or of equipment . . . When one considers the chain of events leading up to the Chernobyl accident, why one person behaved in such a way and why another person behaved in another, etc., it is impossible to find a single culprit, a single initiator of events, because it was like a closed circle.[22]

Rather than place the blame squarely on operators and managers, his seemed a more measured assessment of what led to the Chernobyl (in Ukraine, the former Soviet Union, where Chernobyl is located, "Chornobyl" is used) tragedy, the world's worst nuclear disaster. Factors were multiple, including reactor design, equipment issues, human error, mismanagement, and the systemic lack of an adequate safety culture sustained through earlier USSR politics.

Two years after the meeting, physically weakened after exposure to radiation at the plant and demoralized by what he felt was a systemic lack of interest in safety precautions, Legasov committed suicide.[23] His commitment to truth and integrity, however, continue to challenge official rhetoric regarding the disaster's root causes.

The emperor is essentially politically irrelevant. Yet many Japanese still consider him a national icon representing Japanese tradition and a symbol of the country's historic resilience in the face of disaster and defeat. It must have been both somber and assuring for Japanese to view Emperor Akihito on March 16 who appeared briefly to offer a taped speech. His message resonated with that of his father Hirohito on August 15, 1945, when Hirohito told the Japanese people that they had no choice but to surrender and "endure the unendurable." Akihito's message was different. His was not an issue to surrender but to hold fast, maintain hope, and reach out and "treat each other with compassion."[24] Akihito's brief plea to his people was hardly defeatist, but a prayer for steadfast faith and hope. "Please do not abandon hope."

Nuclear Free?

Global ripple effects illustrate our inescapable in-between, our *aidagara*. Just three months after the disaster, Germany announced its commitment to replace its reliance on nuclear power to renewable sources by 2022. At that same time, Switzerland and Italy decided to turn more exclusively to renewable energy resources. Japan's reprieve was short-lived, however, as in May 2012 government officials approved reopening reactors. This sparked massive protests throughout Japan, yet the government held its line on nuclear reliance. And despite large protests, mainstream media gave little notice. Japan's most popular newspaper is *Yomiuri Shimbun*. The newspaper gave the July 2012 anti-nuclear demonstration of 170,000 little to no press. This concretized all the more the silent marriage between the government, nuclear industry, and mainstream media.[25]

The Chinese ideogram for "crisis" consists of two characters: "problem" and "opportunity." Though horrifying, the crisis erupting at the nuclear complex presented an opportunity to rethink Japan's long reliance on nuclear power. Naoto Kan quickly shut down all plants. This did not sit well with the powerful "nuclear village," the perverse alliance of industry, lobbyists, government officials, vocal nuclear experts, and mainstream media. However, public protests clearly demonstrated strong public sentiment against nuclear dependency. Over 60,000 protestors gathered in September, 2011, in Tokyo. For such a large demonstration, mainstream media coverage was minimal at best. The banner of protest centered around the "Safety of Our Children." With Japan's future embodied in children, this protest further demonstrated Japan's precarity as well as ours—that none of us any longer live in a safe and secure world.

For a brief period of time that first year following the events of 3.11, all of Japan's nuclear reactors were offline, temporarily shut down. By May 6, 2012, none were operating. Japan was nuclear-free for the first time in decades. This freedom, however, is short-lived. Two months later, July 2012, two Ōi reactors at Fukui Prefecture restarted. After Democratic Party of Japan (DPJ) leader Naoto Kan resigned and was replaced by the same party's Yoshihiko Noda as prime minister, who was more sympathetic to the nuclear establishment, successive government leaders faced enough pressure from the powerful "nuclear village" and related industries to restart nuclear reactors. Clearly, the move was not only political but in the interest of Japan's failing economy, which would drop even more dismally and suffer monumental losses in capital from disabling the country's nuclear dependence which enabled it to be a world leader in nuclear power. With the opposing Liberal Democratic Party's (LDP) landslide victory in December 2012, Shinzō Abe is now Japan's prime minister, and one who more clearly favors nuclear energy. Up until 3.11, Japan was the world's third largest producer of nuclear energy. The major political wave is going in the direction of enhanced nuclear energy, despite public opinion polls indicating a desire to be nuclear-free, *datsugenpatsu*. This desire is evidenced by a collective will to conserve power that went well beyond the national 15 percent reduction goal.[26] Any hopes for a nuclear-free Japan were drifting away into a far-fetched utopian dream.

Blame Game

By now, it should be clear to the reader that March 11's calamity in its impact, unfolding, and aftermath is extraordinarily multidimensional. Much too complex to label it as either a natural disaster or human-caused. Because the 3.11 tragedy is so byzantine, we, perhaps out of our need to somewhat ease our understanding of the event's enormity, tend to reduce and simplify it well out of proportion, in order to "provide solace and restore some cognitive order."[27] This must be kept in mind with my encounters and conversations with survivors. Trauma often leads to memory gaps, so that further simplification helps frame the narrative and aids in understanding and sharing. Yet simplification also distorts, exaggerates, and maims the reality. This is apparent when we refer to the sequence of earthquake, tsunami, and nuclear meltdown as merely a "natural disaster." Japanese may refer to the earthquake and tsunami as *tensai*, or "heavenly disaster," similar to what we call an "act of God." Whereas thinking of it as *jinsai*, "human-incurred disaster," as in the case of the nuclear meltdown, in effect discolors the enormity of the disaster.[28] At issue here is the matter of cause. Cause can refer to multiple forces from the more immediate to those deep-rooted. Curiously, the all-powerful magnitude 9 earthquake itself plays an almost irrelevant role in prevailing narratives, even though it ultimately spawned the ensuing immeasurable ruination. When we seek causation, we attend to more immediate triggers.

The distinction between natural and human-incurred disasters is significant. A so-called natural disaster, *tensai*, an act of God, spurs feelings of helplessness and resignation. It cannot be helped. It is unavoidable. On that account alone it is neither considered unfair, unjust, nor inequitable. Injustice enters in when what is instead avoidable and unfair co-mingle. As for causation, we could single out inhabitants who choose to live close to the sea as complicit in their own destruction, particularly given the Sanriku coastline's tsunami history with giant tsunamis in 1896 and 1933. But can we justifiably assign responsibility? We still do not imagine, or remain incapable of imagining, that it will happen to us, neither defying nor denying history, but rather sidestepping it in order to live in the present.

Jinsai, human–incurred disaster, is more difficult to embrace. We see this just four months after 3.11 in another land of peace, Norway. On July 22, 2011, Anders Behring Breivik exploded a bomb in Oslo's government district killing eight, and then methodically murdered sixty-nine, mostly teens, at the Labor Party's annual summer camp on the tiny island of Utøya, near Oslo. The confessed killer committed Norway's worst violence since World War II. During the long drawn-out ten-week trial that ensued, Are Holen, respected psychiatrist and professor at Trondheim's Norwegian University of Science and Technology, was a key testifier in the trial. In my conversation with him, he remains convinced that the killer is certifiably insane, suffering from psychotic paranoid schizophrenia, a loner who lives in his own concocted universe. Breivik was found guilty. But not on grounds of insanity. He was judged to be sane and guilty on criminal grounds. This was apparently the preference of outspoken political voices. This is an extreme example of *jinsai*. In my July 28, 2012, *Albany Times Union* ethics column, I write further about this Norwegian tragedy.[29]

But *jinsai* occurs in more subtle ways. *Jinsai* is even harder to bear when accompanied by scandals like the shifting and misuse of relief funds. And with *jinsai* and forced evacuations, despite the absence thus far of radiation-related deaths, the blame game intensifies. Whereas *tensai* connotes the unavoidable, *jinsai* clearly evokes human error, manipulation, and malice. As for blame, the usual suspects continue to be Tokyo Electric Power Company's (TEPCO) gross mishandling of the disaster which TEPCO later admitted, the "nuclear village's" faulty communication, and outright corporate greed, all compromising safety and oversight. Truly a man-made disaster, *jinsai*.

Blame can extend from misplaced to outright irrational. Even in this twenty-first century, some have pointed to the 3.11 tragedy as punishment from the gods for rampant hedonism and materialism. One of the most glaring and embarrassing examples of this irrational censuring was the governor of Tokyo, Shintaro Ishihara. Ishihara claimed that Japan's "selfish greed" brought about "divine punishment" in the form of its 3.11 triple disaster.[30] Amazingly, citizens reelected him for another term just months after the disaster. Blame also rained down upon foreign journalists whose accounts were accused of nasty hyperbole and exaggeration. Numerous critics have also blamed Japan's mainstream newspapers who underplayed the immensity of the calamity. As for stretching blame, what about the communities who embraced the Daiichi Nuclear Power Plant as their primary employer? What about those vulnerable and poor communities willing to trade their quality of living for a job?[31] Seduction works both ways. Particularly those desperate enough to do whatever they can to eke out a day-to-day existence and have a job, steady income, family stability, and food on the table.

PURGATORY

Survivors of the earthquake and tsunami as well as residents in the vicinity of the Daiichi nuclear complex who were not damaged by Nature's force fled their homes and continue to face additional anguish. They are reeling not only from the magnitude 9 earthquake and colossal tsunami, but from the shock of nuclear meltdown and its ongoing threat of contaminated food, water, milk, farm produce, rice, vegetables, fish, meat, air, ocean, and soil.

Their struggle is further daunting as they have to deal with a top-down, heavily bureaucratic system with complex vested interests. Former residents also face the perverse collusion of politicians, METI (Ministry of Economy, Trade, and Industry) bureaucrats, TEPCO, and other heavily monopolized regional utilities, all part of the "nuclear village."

It is highly unlikely that residents will return to their abandoned homes in places like Odaka. Questions surrounding the use of property in those areas are still not settled. Furthermore, for numerous former residents, given the long-term effects of radiation, the future remains uncertain. And since compensation can no longer be claimed after twenty years, this leaves them in a permanent state of limbo. Evacuees, whether voluntarily leaving or forced to leave, are in their perpetual purgatory. They inhabit this anguishing in-between. This is not purgatory in the Catholic sense, as bathed in traditional teachings regarding the enduring of needed pain and cleansing, sketched in between earthly and eschatological time in order to eventually enter the promised state of perpetual bliss. This religious conceptualizing of in-betweenness represents a significant moment in intellectual history, what gifted medievalist Jacques Le Goff calls the "spatialization of thought" to describe our mental visualizing of a purgatory.[32] In similar fashion, Watsuji Tetsuro offers a spatializing of time, in that our living here and now is not separate nor siloed, but rather imbedded in in-betweenness. *Aida* owns no territory. Its only meaning lies in its relatedness to something other but not apart. For post-disaster evacuees, particularly for those displaced because of nuclear radiation, there is no future they can look forward to. They inhabit a futureless diaspora. They are lost in the notches of time, in-between on the one hand a past of promises, dreams to be fulfilled, the usual banalities and new awakenings, and on the other hand a future that is left blank.

Home is not simply some space where one lives. Place matters. For Hamako Watanabe, home meant living in her home in Kawamata, Fukushima Prefecture, and working together with her husband on a small chicken farm. One month after the nuclear disaster at Daiichi Nuclear Power complex, her home, forty kilometers, nearly 25 miles, away from the plant, and the surrounding area were declared an evacuation zone. She and her family were forced to leave, and they fled to an apartment in Fukushima City. The chicken farm, her livelihood, closed down. Since then, she became increasingly depressed and not able to see any light at the end of the tunnel. She lost all hope for returning home. After deep depression, in July 2011, fifty-eight-year-old Hamako went to her abandoned home, soaked herself in gasoline and set herself on fire.

Her husband, Mikio Watanabe, later brought civil suit against TEPCO, charging them with responsibility for his wife's depression and suicide. Earlier claims had been resolved through out-of-court settlements. But the result in this case was different. In a landmark decision this past August, 2014, Fukushima's District Court ruled that TEPCO, Tokyo Electric Power Company, pay ¥49 million yen (around $471,000) in damages to Hamako's family, declaring that the power company was responsible for her eventual depression and suicide. Though the plaintiffs requested nearly twice as much, this was the first time that a court ruled regarding compensation over a 3.11 nuclear-disaster-related suicide.[33] According to the ruling, "It is well assumed that the stress caused by sudden loss of the base of her life against her will and unknown future in evacuation was unbearable for her."[34]

The evacuation and diaspora of "nuclear communities," towns seriously affected by the 3.11 nuclear meltdown at Fukushima's Daiichi Nuclear Power Plant, essentially means that in all likelihood former inhabitants may never go back. Nearly 136,000 Fukushima residents are still displaced throughout the prefecture. The total throughout Tohoku is 260,000 *kasetsu* dwellers. Current figures now list more deaths in Fukushima from stress and mental illness than from the disaster itself. Deaths from stress-related illnesses (physical and mental stress, depression, and suicide) total 1,656 Fukushima people. Disaster-related deaths number 1,607. Most of these "stress deaths" are of those over sixty-six years-old.[35] This means that in Fukushima, following the Daiichi nuclear disaster, the most critical long-term issue concerns the mental health of residents who still live in temporary shelters. In a relatively large-scale study of *kasetsu* residents in Fukushima Prefecture, researchers have established a direct link among unemployment, physical and mental stress, family stress, social capital, social support, and suicidal thoughts.[36]

This diaspora compels us again to underscore the vital imperative of *aida* and the ensuing trauma for those whose connectedness is now shattered. Insults to our *aida* are compounded when such rending of communities is compulsory. Moving the community in this case means dispersing it. When enforced, it violates our natural in-between.

Tokyo's Colony

Survivors' anguish is further deepened due to Tohoku's history, which we described earlier. Tohoku has been alluded to as Tokyo's "colony." To begin with, Sanriku's thriving fishing industry, especially sardines and flounders, has long served the population of Tokyo and other major cities more than its own. And the Daiichi power plant serves Tokyo and the surrounding metropolitan area, not the people of Tohoku. Still, the plant provided rich opportunities for employment for locals.

This service to the capital becomes especially strident in the case of TEPCO's compensation to hundreds of thousands of citizens displaced from the nuclear meltdown. Not only has compensation been meager (less than

$20,000 per victim over a year after the disaster,[37] but the bureaucratic process of filling out forms to prove victimhood has been bogged down with bureaucratic ineptitude and red tape galore. This incredibly inefficient attempt to prove victimhood with hard evidence, like the ludicrous need for official receipts when city offices and records had been wiped out by the tsunami, comes close to resembling the farce after the Bhopal, India, disaster. Adding insult to injury, TEPCO would only offer compensation if those who evacuated were officially required to do so. Only those required to leave could claim damages. Given the mixed messages regarding radiation risk, communication lapses among TEPCO officials, town leaders, and the government, and understandably widespread fears that children might be contaminated, those who "voluntarily" evacuated are not entitled to claim compensation.[38]

In effect, many in Tohoku had been victimized well before March 11, 2011. Not only in Tohoku, but in other areas that house nuclear plants. Historian Kawanishi Hidemichi refers to these with the sociological term "national sacrifice zones." According to Hidemichi, Tohoku is, in effect, "Japan's nuclear power colony."[39] This leads us to question the wisdom of having these nuclear power plants in such volatile spots and in a country that regularly undergoes large-scale earthquakes. Japan experiences nearly 20 percent of global earthquakes with magnitude 6 and over. The selection of these nuclear plant sites raises a host of questions. Situated directly on fault lines and shifting tectonic plates, for instance, leaves Shizuoka Prefecture's Hamaoka five-reactor plant, only 113 miles from Tokyo, especially vulnerable.[40]

Purdue University political scientist Daniel Aldrich has conducted exceptional research on this issue. He argues persuasively that choosing specific sites involves choosing communities experiencing economic hardships and at the same time lacking strong social bonds to sufficiently protest setting up sites near their communities.[41] The nuclear establishment had been targeting vulnerable rural areas with declining population and economy as sites for nuclear plants while touting safety and promising an economic lifeline—jobs and money. The strength of civil society and community, and thus the degree of social vulnerability, often resides in grass-roots organizations and collectives of, for example, fishermen, farmers, and neighborhood associations. This is similar to the notion of "social capital," the seeds of which trace back to Alexis de Tocqueville and are described at length in Pierre Bourdieu's *Forms of Capital* and more recently in Robert Putnam's *Bowling Alone*.[42] With this in mind, note that nuclear power plants in Japan are often situated on demographic, social, and economic fault-lines. They are deliberately placed in settings where local residents are already marginalized and thereby powerless to resist the lure of jobs. Like Fukushima Nuclear Power Plant I and II, these plants suture residents to job security. Due to local residents' economic fragility, the poorer populations tend to become less outspoken in resisting the lure of money and jobs. This weakens a community's capacity to mobilize and oppose government hegemony. Fukushima Daiichi Power Plant is located in coastal Futoba and Ōkuma. When the plant was commissioned in 1971, the region had been facing a waning economy particularly in Ōkuma with the dwindling power of fishing cooperatives and farmers' groups. Thus their region's "social capital" remains in jeopardy. This painfully demonstrates *aidagara* in the face of economic, political, and institutional hurdles. However, with the younger generation, with students like Tomomi from Fukushima, whom I will soon describe, and many other volunteers with whom I worked in Tono, perhaps this will change as their protest is simmering.

Hibakusha—The Fourth Disaster

Disasters generate unfounded fears. Unless countered immediately by solid science and facts, these fears become infectious, spreading among the uninformed populace, spawning a culture of fear. The nuclear meltdown on 3.11 detonated a fourth disaster—the stigmatizing of all people and things associated with Fukushima Prefecture, home to the devastated Daiichi Nuclear Power Plant.

Here, there is no describing the tragic irony comparing the fate of displaced victims of the Daiichi plant's nuclear fallout and those victims of the human-imposed Hiroshima and Nagasaki's nuclear calamity. The death count from the Pacific War's atomic bombings easily surpassed 225,000. Those who survived were called *hibakusha* (bomb blast people), and, for many of them, survival may have been a fate worse than death as they faced enduring discrimination based on ungrounded fears that their contamination would be contagious. The science of radiation and its long-term effects on humans was in its early phases. There was at that time little understanding of the far-reaching impact on future generations of survivors.[43] But even now, seven decades later when the science of radiation and its long-term effects on humans has gained more solid ground, public fears and stigmatization continue unabated, particularly among many Japanese, who should know better, given their tragic personal history.

After the August 1945 bombings, the names for "Hiroshima" and "Nagasaki" were written in katakana instead of the conventional Chinese kanji. This distinguished the cities of Hiroshima and Nagasaki (normally written in the formal kanji) from the ignominious and shameful "event" (written in katakana). The word "Fukushima," also customarily scribed in Chinese kanji, 福島, is now sometimes written with the Japanese syllables of katakana, フクシマ, distinguishing it as the site of disaster instead of a prefecture or city, just as "Chernobyl" and "Bhopal" elicit the sense of calamity. This leads some activists who protest this discrimination, or simplistic way of thinking about "Fukushima," to write "Fukushima" in hiragana, another Japanese system of syllables, in order to avoid bias.[44] Japanese children first learn hiragana as it is informal and conveys a sense of ease. Only after they have mastered hiragana, do they then learn katakana and kanji in that order. Therefore a place name such as "Fukushima" when written in hiragana is less formal, more relaxed. My friend Kunio tells me that someone may deliberately use hiragana more as a point of emphasis, rather than, in the case of "Hiroshima," an outright protest. As in everything, especially in Japan, so much depends upon the context. The hiragana for "Fukushima" is ふくしま.[45]

In reality, no matter where they go, survivors, especially young families now often face discrimination for the simple fact that they are from Fukushima. Amidst widespread, unfounded rumors regarding radiation spewing forth all throughout Fukushima and now through other parts of Japan, these displaced survivors from towns in Fukushima Prefecture continue to face bias. They are the new *hibakusha*. Sadly, children are not immune to this stigma. Many mothers would prepare their children's own lunch fearing that school lunches would contain contaminated food. Theirs fears are grounded in realities and reports they've closely followed. High levels of radioactive cesium had been found in the hay fed to cattle before cattle were shipped to markets for meat. And popular vegetables like spinach, bamboo shoots, and mushrooms were also found with higher than normal amounts of cesium. These vegetables were shipped over two hundred miles away. In any case, protecting their children meant that their children would also be noticed. Standing apart from the crowd is often viewed with suspicion among many Japanese. Children especially fear being ostracized and made fun of by peers. Children bringing in their own boxed lunch incur inordinate teasing from classmates. Even teachers got in on the act by berating these poor kids.[46]

Kyoto holds an annual bonfire on the side of a hill overlooking the city for all to view. Much wood is used for burning. Since 3.11, after protests from some Kyoto residents, organizers of the bonfire have refused to use any wood from Fukushima Prefecture, fearing that some of it may be contaminated with radiation. There is even fear of using wood from Sanriku, beyond Fukushima. Whether there is indeed scientific evidence that the wood may be contaminated is not the point. Only minimal protest from citizens of Kyoto, who are generally more affluent than people in Tohoku, was enough to raise the question of even possible contamination. The same protest occurred in the "charity aid and hope" fireworks festival in Aichi Prefecture, east of Kyoto. Here, organizers refused to use Fukushima-made fireworks. These are rituals of hope. Yet, wood from the Fukushima region is rejected. Will the real disaster stand up? Again, this gives us pause regarding generalizations of some ideal image regarding Japan's collectivism and community identity. Here we have a deeper social, communal, "soul" disaster.

Shima kuni kanjō literally means "island country emotion."[47] This is the small-mindedness that we find with parochialism, the parochialism that comes about from the mind-set that one's world is one's island, and nothing meaningful beyond. This is the dark side of a collectivist mentality and society that unquestioningly conforms to the group. Volunteers with whom I worked came from all over Japan. Two students from my birthplace Fukuoka, Kyoko and Ayaki, talked about this mentality and wanted to be free from this way of thinking. It occurs when locals are so enwrapped in their own worldviews and ways of looking at life, they fail to look beyond their own "island" or borders. They know no other world than their own, and assume that their own beliefs and values are all that matters. This way of thinking sparks the "not in my backyard" (NIMBY) attitude in communities who oppose anything that interferes with their lifestyle, like the NIMBY mind-set at work here in the United States with respect to having "undesirables" like mentally disabled persons, prisoners, welfare recipients, and waste plants as neighbors.

MINAMISOMA

Our journey to Minamisoma was one of our starkest days. I was with East Japan Railway Cultural Foundation's Kunio, Mizuki, Noriko Yamamoto, and our driver Kushida. The cold and penetrating wind reminded us of the invisible, and insidious threat of radiation. Due to winds, radiation was carried from the Daiichi plant to neighboring mountain villages through which we drove. Some of the mountain villages are nearly entirely abandoned, their citizens forced to evacuate or who evacuated voluntarily. Many left voluntarily out of fears of radiation for their families. This has split up numerous families as many elderly have also chosen to stay in their own homes, regardless of the threat of contamination. The sense of place and home is more deeply imbedded in older generations. In contrast, young parents are fearful for the safety of their children, thus leave these towns, leave their aging parents, and seek work and life elsewhere.

En route, we drove by Iitate, now abandoned ever since the enforced evacuation in March after the nuclear reactor explosions. The town is situated in stunningly beautiful wooded mountains, a sharp contrast to the invisible radiation that permeated the area. Here, along with Okuma, there were exceptionally high traces of these toxins. Children, closer to the ground, were especially vulnerable. Imagine what it is like to be forced to evacuate your home, your dwelling, with all of its still life and memories. Though different from a Krystalnacht where circumstances were churned out by human malice, in the case of Fukushima, natural "malignant" forces combine with human hubris and blunder. And all life in the mountains was adversely affected by radiation levels, including wildlife, birds, trees, foliage, and plants. Residents here were told to leave immediately. Picked up by busses, they had to leave behind their cars. The only ones now in the village are police, keeping a watchful eye patrolling the area ensuring that no one stays overnight. Residents are occasionally permitted to enter in the daytime to check up on their homes and belongings. Farmers would sometimes come and check their land. For many of them, their cattle were left behind and starved to death. Before disaster struck, many residents only moved to Iitate recently and built new homes, seeing it as a beautiful, quiet place to retire. Now, they're gone and still pay mortgage on a home they no longer live in.

When I volunteered during the winter of 2014, almost all of the twenty-two million tons of debris were removed, and the landscape was one barren flatland, cleaned but empty. Even the debris along the Fukushima coast was eventually removed, though only after much resistance due to fears that the rubble would be contaminated with radiation.

Just think of the monumental task of scraping away contaminated soil. The government requires that two inches of topsoil must be removed, even though many farmers contest that, claiming that the radiation has penetrated much deeper. As Watsuji reminds us, climate will always have its sway. Remember—we are inescapably in-between. So, with more rain and snow that season, contamination can sink even deeper. Although the crucial need for this has spurred on local construction industries and companies as far away as Tokyo, this has resulted in numerous piles of waste. Now imagine the incredible amount of contaminated soil cleared away and placed in large black bags. These bags are spread throughout the countryside. So, where should these large black bags with contaminated soil be placed? Buried? This is part of the future legacy of horror and uncertainty that area residents face. It's no wonder that countless families have moved away, far away.

Even though the mountains of rubble had been removed, city officials still face the daunting question—Where do we put it? Bury it deep into the ground, knowing that the half-life of radioactive cesium would threaten far into the future? The principal source of radiation from the Chernobyl disaster was cesium-137. Amounts of cesium-137 and cesium-134 have been detected from the Daiichi plant as well. Given cesium-137's half-life of approximately thirty years, this means that in thirty years half of the atoms will have decayed. Of the amount left, half of that will decay in another thirty years, and so on. Cesium-134 has a half-life of around two years. What about the 200,000 tons of contaminated water at the Daiichi plant? This water is stored in one thousand ton water tanks. Where will this contaminated water go? The spread of radiation from the plant has already reached out to 8 percent of the country.[48] Moreover, the medical effects of the radiation remain a hotly contested issue, and this further intensifies the prevailing tone of uncertainty among those who live in Fukushima Prefecture.

Experts now claim that radiation levels, for example from cesium, are safe enough in Minamisoma. But tell that to parents. The fear level remains high. After all, how can you trust official assurances when the government raised the maximum permitted limit of radiation exposure per person from one millisievert per year to a whopping twenty millisieverts just months after 3.11, in an effort to prevent mass panic, hysteria, and evacuation?[49]

Mayor Katsunobu Sakurai

When we arrived in Minamisoma, its long-term damage from the earthquake and tsunami was apparent. Like many towns along the coast, its devastation is incalculable. This southern town in Fukushima Prefecture deals more intimately with the threefold tragedy of earthquake, tsunami, and nuclear meltdown. The earthquake and tsunami combined to take 947 citizens. Of these, nearly one hundred were children. Dread of radiation spurred over one-third of the population to leave the city. On March 12, citizens began their mass exodus, traffic flooding the roads for days. Out of a population of around 70,000, 27,000 have left, including a sizable portion of city officials. They may never return. All this categorically affects the city's life expectancy.

Minamisoma City, in Fukushima Prefecture, is a recently merged municipality of three different cities—Haramachi, Odaka, and Kashima. Some experts view this merger as negatively impacting Minamisoma's sense of identity and community. Aldrich and Sawada point out that new municipal mergers have generally experienced higher death rates than those in long-established, more historically autonomous municipalities.[50] Minamisoma's merged status therefore offers a further challenge to Mayor Sakurai to ensure sound coordination of evacuation and mitigation efforts. This is no easy challenge for the mayor, particularly given long historical tension within the various regions of Fukushima Prefecture.

Fukushima is the third largest of Japan's forty-seven prefectures and has three regions: eastern coast, middle, and western mountains. Its Pacific coastline stretches nearly one hundred miles. Since the restricted area around the Daiichi Nuclear Plant takes up less than 10 percent of Fukushima's total area, most of Fukushima is now considered safe to inhabit and visit. The mayor's challenges are especially acute given Fukushima's internal tensions dating back to the Edo Period (1603–1868). Towards the end of Tokugawa rule, the growing influence of Western ideas and Western weaponry challenged the authority of the shogun. While many clans sided with restoring imperial rule in the eventually new Meiji government, the Aizu clan remained loyal to the shogun, Japan's last, Yoshinobu Tokugawa (1837–1913). This led to a short-lived, bitter conflict, the Boshin War, from 1868 to 1869 in which the Aizu clan with a few others were defeated by Meiji government forces. The emperor supported the new government and changed the name "Edo" to "Tokyo." The Aizu clan received no help from those in Fukushima in both middle and coastal regions. Since then, relationships among the three regions have been strained. Makiko Okuyama, head of Psychosocial Medicine at Tokyo's National Center for Child Health and Development, spoke to me of her friend who comes from the Aizu area. People there claim they would not marry anyone from other areas in Fukushima, especially the coastal areas. For Makiko, the Aizu area is still stuck in the past. She points out "For us, the last war was World War II. For those in Aizu, the last war was the Boshin Senso [Boshin War] at the end of the Edo period."[51] In any case, Mayor Sakurai faces a difficult challenge in integrating the three regions.

In the spirit of his samurai ancestors, Katsunobu Sakurai, Mayor of Minamisoma City, is unflinching in his commitment to rebuild his community. Residents there and throughout Fukushima Prefecture are not only earthquake and tsunami victims, but suffer from the ongoing, invisible threat of radiation poisoning. They also endure that "fourth disaster" of unfounded rumors and stigma attached to this threat. For instance, in nearby Odaka, a father had been using his son's car prior to 3.11. The father was a good friend of Takada Motoyuki, our driver throughout our visit. After the disaster, his son pleaded with him not to return the car. License plates carry the names of the prefecture where cars are registered. There are abundant rumors that people in Tokyo have treated cars from Fukushima differently.

During my meeting with the small, compact, and confident Mayor Sakurai, he stressed in measured fashion a critical need for continued research into and application of alternative renewable energy sources, like solar and wind. As for research into technologies for geothermal and solar energy, Japan is one of the world leaders. Nearly every volunteer I worked with felt the same way. Many of them were students who switched their majors from just about all things related to nuclear studies to majors in fields like social work, health sciences, literature, and law. And almost all of the survivors I spoke with were steadfastly opposed to Japan's fixation on nuclear energy.

Without any doubt, the mayor faces formidable challenges then and now in his ongoing efforts to rebuild his community. As a start, throughout the unfolding dismemberment and debacle at the nuclear site, the mayor had little official information at his disposal. Any information that was available was both scattered and appeared to play down the nuclear threat, supported by commentaries from industry and pronuclear experts. Both government and TEPCO kept him in the dark so that he had little data to work with and share with his citizens. It took three weeks before TEPCO officials finally shared with him the dire situation at the Daiichi plant. He and his citizens also faced serious lack of gasoline, food, supplies, healthcare, and news coverage. Gasoline tank drivers, fearing contamination, did not come near the town. Gasoline was at its ebb as most citizens headed for the hills with their families. The fear that radiation poisoning had taken over Minamisoma in effect one-sidedly quarantined the town. Like the villagers in the Algerian city of Oran in Albert Camus's *Le Peste* (The Plague), the remaining citizens of Minamisoma suffered the indignity of isolation and the stigma of being victims. As with the gasoline tank drivers, delivery truckers of food, water, and supplies also shied away from entering the area. As for those who remained in the area, including elderly, disabled, and bedridden, they faced the threat of starving. Healthcare professionals also took part in the town's mass exodus, leaving many bedridden patients on their own. There is no lesson in community here. Nonetheless, we must be careful of generalizing. Many caregivers did remain by their patients, refusing to abandon them. Whereas media accounts often portrayed doctors and nurses abandoning their patients, those who decided to be with them and care for them, like those at Otsuchi Hospital, did not make headlines. Doctors and nurses at Otsuchi Hospital died along with their patients as waves consumed them.

A special problem for the mayor was that people beyond Minamisoma and the nuclear site exclusion area, indeed the rest of Japan, were in the dark about the city's plight. Mainstream journalists were ordered by their companies to leave Minamisoma, typical of mainstream media companies' responsibility to ensure the safety of their employees. However, this meant that the plight of the town was not reported to the rest of the world. Those remaining in Minamisoma were therefore exiled, isolated, and abandoned. The mayor therefore took it upon himself to deliver a public plea before a camcorder set up in his office, an eleven-minute appeal that he uploaded to YouTube with English subtitles.[52] This was his plea to the world for help. "I beg you as the mayor of Minami-soma city, to help us." His appeal reached millions worldwide. That and his steadfast efforts to save and inspire hope in his citizens later earned him a place in *Time Magazine*'s 2011 Top 100 Influential People. As with my interview with Tono mayor Toshiaki Honda, here is a brief summary of my lengthy conversation with Mayor Katsunobu Sakurai.[53]

> Katsunobu Sakurai (KS): While the government hopes to restart the Daiichi nuclear plant, it is crucial that officials make every effort to truly comprehend the reality of the monumental and far-reaching damage here. Most importantly, they must understand the suffering of our people. All of us continue to suffer in many ways: parents, elderly, all those in shelters, and so forth. I know their suffering. It continues. The disaster is far from over. The accident at the power plant raises long-term concerns about everyone's health. Young women wonder whether they will have healthy babies. Many elderly who live in

kasetsu feel trapped and do not want to live their remaining lives there. People who live in *kasetsu* keep asking me to help. Mentally and physically, they are reaching their limit.

Officials must understand the situation. Since Tokyo is around 250 kilometers (155 miles) away, many people may think they are safe. They say that the people of Fukushima are resilient and will endure. And there is financial aid, even though people who voluntarily evacuated do not receive money. But money by itself is not enough. Our educational environment has been torn apart. Children and parents face many issues having to go to other schools outside their community. I hope that in my position as mayor I can work with others to better understand our crisis. If the government wishes to restart nuclear plants, they need to also do their best to help us all recover. Many who live in temporary shelters and elsewhere feel that the government and the nuclear company can do much more. What leaders need to do is give hope to people. We are not against reopening the nuclear power plant. However, we should first work to recover our area, and then restart the plant.

As this city's mayor, it is my duty to help government leaders better understand the reality from our point of view. I am chief of the committee of Soma Nomaoi [the annual samurai and horse festival]. Before the shrine I pray that I stay mentally strong. "I have to stay strong." During 3.11, I could not lose my focus. I needed information to make crucial decisions. Yet, on March 11, everyone was in a panic. And we were without adequate information. On 24 March, I had to evacuate our residents. Our citizens evacuated to 120 places around the country. I then sent 120 employees to these places. I also sent nurses. Evacuees felt safer with these nurses, and these nurses stayed with evacuees for around three months.

In our city, we lost 636 citizens to the tsunami. Over eighty are still missing. When I jog daily in the bay area, I feel their suffering. To all those who are alive and still suffering I tell them they are lucky—they are still alive. "Appreciate the fact that you are alive. Make the most of your life. Don't think of money. Though we say we have lost everything, we still have tradition, and we live in tradition."

MB (author): We Americans have much to learn from all this. Your people who continue to suffer need to have a future to hope for. This tradition that you speak of, Soma Nomaoi, which is over one thousand years old—do you think it offers support and hope to people?

KS: Definitely. People here are very strong. I myself will participate in the festival because my family has supported me. In fact, I, my parents, my grandparents, and their parents have traditionally participated.[54] Many here who have lost family will not participate. Yet many others will because their families had participated in the past. There are many from Minamisoma who have evacuated to Fukui Prefecture. They have shared with the people there the history of Soma Nomaoi. "Nomaoi is part of their life; it is not just a once-a-year festival, but is their life."

When I asked the mayor his primary message to the world, he responded that we must realize that, even though the disaster is technically over, it still unfolds in the suffering of countless survivors. Can we grasp the reality of this real suffering, of those displaced in *kasetsu*? "Of elderly who fear dying in these shelters?" Like Mayor Honda, he devotes his life and work to survivors' recovery. Only in this way can we resurrect community. His final words: "What is most important for me is to work for my citizens. All I can do is work for my citizens."[55]

Odaka, Nuclear Ghost Town

Most of the houses in Odaka are intact, but uninhabitable. Their owners cannot live here. Odaka is one of many "nuclear ghost towns" and is situated on the edge of Minamisoma and near the rim of the twelve-mile exclusion zone, the perimeter around the Daiichi Nuclear Power Plant where no one is allowed entry other than officials, plant workers, and clean-up crews. Here, silence speaks loudly. The entire town is empty—the fire department, community center, train station. Figure 7.1 shows Odaka's main street, three years later.

Citizens were told to leave right away. There was no time to pack. They had to leave behind their cars. Students had to immediately board busses heading out of town. Their bicycles are still where they left them that morning.

Once a key and thriving Joban Line stop in East Japan Railway's powerful rail system, Odaka Train station is now filled with weeds. Numerous birds nest in the abandoned station, their chirping full of life in contrast to the eerie stillness of the deserted town. Its rusting tracks are sixteen kilometers (nearly ten miles) from the power plant. There's an empty pig farm nearby with rusted pigsties. When we walked through the station, there was still water in the building, situated two kilometers, just over one mile, from the ocean.

Takada Motoyuki, in his early forties, lives in Odaka, or used to until 3.11. Takada, the former Japan Railway East Haranomachi station master at Odaka, generously offered to be our driver throughout our visit to Minamisoma and the surrounding areas. He was taking a computer class when the earthquake struck. After the shaking stopped, the class was immediately dismissed. He knew a tsunami would soon arrive. Within minutes, he sped home. He and his son spotted the oncoming wave, around eight hundred meters away, from their second floor. It looked to be around fifteen meters, or 49 feet, high. They fled immediately, escaping safely to higher ground. So did most of the town. Many in the vicinity immediately jumped into their cars heading towards higher elevations. At one point, one of the cars on the road stalled. Even after a neighbor in the car behind tried to help, the car would not start, creating a terrible traffic jam. In seconds, numerous people and their families were killed. Just past his neighborhood is flat land and rice paddies. Waves swept through his home, his neighborhood, rice paddies and

Figure 7.1. Nuclear ghost town, Odaka, 2014

all surrounding farmland while electrical poles toppled. He had grown his own vegetables, but all that is now destroyed. Of the hundreds of houses in his area, sixty-six were left standing. fifty-four people were killed, and 181 are still missing. Even neighbors on slightly higher land, thinking they were safe, were swept away, including his parents and one grandparent who lived in a separate house. These were relatively newer homes, around thirty years old.

Since impact, Takada, his son, a grandparent and grandchild stay at an evacuation center. Weeks later, once authorities deemed it "safe" enough to enter the area, he returned to his drenched home to retrieve a stove for the evacuation site. Now, he and others are prohibited from going to their homes. And due to radiation levels, he could never sell his property. I asked Takada whether owners still pay mortgage on their homes even though they cannot return to them. Takada responded that they do. If they find a place to build a new house, they will pay two mortgages, unless they can cancel the first loan, which is unlikely. There are talks about resolving this double-mortgage issue, but to no avail. Even now, much rubble still remains including remnants of houses, cars, trucks, and tons of other debris. Much of this resembles the rubble we witnessed two years ago when we visited towns like Rikuzentakata, Otsuchi, and Kamaishi further north in winter 2012. Debris still remains because of fears of radiation. Takada once came across bone pieces. Because the soil remains wet in the lowland, body parts may have sunk deeper into the damp ground. While we were there, we spotted poles with white flags placed there by the SDF. These marked sections where the SDF, after searching the three obligatory times for remains, no longer found anything.

We also noticed the strong presence of police. Police from all over Japan were here. According to Takada, after the waters receded, there were some "despicable people" who came here to steal from bodies—jewelry and other valuables. Once more, defying our image of a harmonious, benign, collective Japan. Thieves in trucks, disguised as maintenance workers, have also stolen cash registers. This is the dark side of our *aidagara,* our in-betweeneness, human parasitic behavior especially in the wake of tragedy.

Another dark side of our *aida* is that there is growing alcoholism among many who feel imprisoned in their own homes, without transportation, and isolated. We spoke with a friend of Takada's who, along with Takada, regularly volunteers to help those who live in *kasetsu*. These *kasetsu* are built outside of the initial thirty mile exclusion zone first established. Volunteers will often hold tea parties to help build community spirit for these displaced persons. At first, very few took part in these, but over the years, more participate, particularly women and children. The real concern lies with those who do not participate. Here is the interior, mental fallout from the disaster.

Minamisoma and its surrounding area have a tough time dealing with mental health issues. Nurses in evacuation centers are particularly sensitive to this. Men tend to gamble. Among men in *kasetsu* alcoholism is especially high. Women can deal with stress by being more sociable. Men, socially conditioned to be more introverted, are better off with hobbies. Men have little interest in therapy, support groups, and counseling, and, according to Takada, they lack the "courage to talk with new people and create new relationships." Because of this interior fallout, while the government was not at first discerning in allocating housing in temporary shelters and crammed everyone together, it is now more judicious and aware of the need to have families and neighbors together. Takada and his friend tell us that in their work in *kasetsu,* they felt as if they were "human rights counselors for *kasetsu* people."

Fukushima's samurai past contributes to how survivors, especially men, deal with trauma from 3.11, particularly the threat of nuclear contamination. Samurai and, in general, Japanese stoicism is such that survivors tend to resist counseling. Though survivors clearly show symptoms of post-traumatic stress syndrome (PTSD), many resist psychiatric intervention. Makiko Okuyama stressed this in our meeting. She believes that with their strong samurai tradition of stoicism, Fukushima survivors tend to deny their condition of trauma. Many feel a sense of shame associated with anything that smacks of psychological disorders or mental health issues. In this respect, Fukushima mirrors much of Japanese culture and history. There has always been a historic stigma attached to mental illness. Though this has slowly changed over the years, there remain pockets of denial and resistance driven by this shame.

MORAL GRIT AND MOVING FORWARD

When it comes to past trauma of major proportions, overlooking or downplaying events are ways of sanitizing history, which past Japanese government and leaders have been known to do with respect to, for example, its recent war history like the shameful and brutal 1937 Rape of Nanking. To this day, Japanese history books have a way of downplaying or ignoring this ignominy. We Americans are not exempt from sanitizing history. Consider the heated controversy in 1995 surrounding exhibiting the B-29 that dropped the atom bomb on Hiroshima August 6, 1945, the controversy imbedded within a persistent culture war pitting left versus right as to the United State's moral culpability in letting loose such nuclear death upon hundreds of thousands of non-combatants. Americans still can't agree upon the moral narrative. Tiny Rhode Island, where our little family settled in Newport after over four years in Japan, is the only state that still celebrates Victory Day, or Victory over Japan Day. This takes place every August, the month when Japan surrendered following the annihilation of Hiroshima and Nagasaki. Legislative efforts to change the name to "Rhode Island Veterans Day" or "Peace and Remembrance Day" have met with rancorous opposition.

Japan can easily look to its recent past as a history of victimization. Its victim sensibility—of its right-wing war mongering military leaders, of an enemy viciously and indiscriminately fire-bombing women, children, and elderly in Tokyo, and deliberate targets of the first and only use of nuclear weapons on humans. And now, citizens remain victims of a "nuclear village" bent on imposing nuclear energy on all of Japan, despite a tragic legacy of nuclear power, of a corrupt and backhanded political-industrial clique, of the lethal mishandling of Daiichi Nuclear Power Plant with its string of communicative faux-pas, of the threat of radiation's invisible poison, and now of victims facing discrimination amidst a dark future uncertainty.

The tragedy of 3.11 can cast a Japanese mind-set further into embracing this role of helpless victim. From what I have experienced and learned speaking with survivors, however, they refuse such an apathetic role and move forward with *gaman*, courage and determination. They resolve to move forward and not look back. They live out their moral grit realizing two things. First, there is no recourse but to move forward. Two, this can only be accomplished together, for they are, like all of us, like it or not, imbedded in *aidagara*, in-betweenness. Only in and through this *aida* can they survive and thrive.

My fellow volunteer Tomomi Kaneko is a living example of this. Tomomi studies civil engineering at Fukushima National College of Technology. I first met Tomomi en route to our work in Otsuchi to clear areas of snow and ice for people in *kasetsu.* We sat together in the van, and she was intensely interested in my reason for being here, asking me point blank "Why are you here?" I tried to explain to her my work and research aim, most especially that of learning from others, like her. Unusually straightforward, she made it clear to me that she and her friends are in fear of having children due to risks of radiation. Her concern was not a matter of ideological opposition to nuclear energy or politics. She was simply expressing a deep human anxiety over its adverse health consequences particularly for children. Despite this anxiety, however, Tomomi is a "repeater" and, though young, aims to become a long-term aid provider, or *chōki*. She aims to use her studies and degree to help rebuild her town where she lives, Iwaki City. At the same time, she will continue her work with Rwandan refugees who have settled in Fukushima, helping them adapt to their new country while maintaining ties to their still brutalized homeland.

Hope, Samurai Matsuri, and Connectedness

As in our earlier discussion of *kagura,* it is clearly the case that celebrating festivals, *matsuri*, sustains the community. They are reminders of cultural history, heritage, and roots. For example, there is the New Year's Daikoku Dance along with various *kagura*. In Fukushima, its annual Soma Nomaoi (wild horses chasing), Minamisoma's most popular *matsuri,* remains its high point. The festival takes place over three days towards the end of July and dates back over one thousand years. It celebrates the region's rich and longstanding samurai heritage. The power of the samurai was especially intense between 1300 and 1600. The area's horse-breeding fame traces back to the legendary Soma clan of samurai who ruled the Minamisoma region and was especially renowned for their warriors' prowess and skill with horses. Tens of thousands of people throughout Japan attend this annual commemoration of samurai values, a festival designated as one of the National Important Intangible Folk Cultural Assets. The festivities are dedicated to gods at three principal local Shinto shrines—Odaka Jinja, Oota Jinja, and Nakamura Jinja.

This brings us back to an earlier question about the role of religion in Japan. Are the Japanese religious? This seems like an odd question as spirituality seems to permeate the culture. Soma Nomaoi shows us that being "religious" is not so much a matter of religious doctrine or dogma. This partially explains why among Japanese questions about the nature of suffering are rare, no stories similar to Job. The 3.11 catastrophe caused widespread death, devastation, and displacement. Yet there were no questions as to why? Existential probing about suffering and meaning is not raised. What brings the community together are the affective elements, expressed through *matsuri* and ritual practices, like the *kagura* dance that brought together thirty or more village survivors near the *kasetsu* in Kamaishi where we worked.

Taking place in Minamisoma and nearby towns, the annual Soma Nomaoi highlights include a spectacular procession of over five hundred warriors and their horses clad in traditional samurai armor (*kachu*) through the streets, twelve samurai clad in armor, helmets, and *katana* (sword) in a heated 1,000 meter horse race (*kacchu-keiba*) with their clan flags on their backs, and all the samurai competing together on a vast field to capture goshinki flags (god flags) shot into the air with fireworks (*shinki-sodatsusen*). The festival ends with *nomagake*, a ceremony in which samurai without armor and wearing white capture wild horses bare-handed. Samurai then present captured horses as offerings in front of the Odaka Shrine.

The Soma Nomaoi celebrates the traditional values of the samurai warriors, more properly called *bushi*, and strengthens the community, reminding it of its cultural identity. The most important samurai values of the "Way of the Warrior" (Bushido) are skill, honor, bravery, loyalty, and, especially relevant after 3.11, courage and resilience in the face of danger, crisis, and death. No doubt, Soma Nomaoi inspires hope among the people who choose to remain there. This is evident in Mayor Sakurai's excitement in describing the July festival. In his famous YouTube appeal to the world for help, on the wall behind the mayor hangs a poster of Soma Nomaoi.[56]

After 3.11, there were no *matsuri*, or festivals. Over one hundred horses and their owners had died in the disaster. Immediately after 3.11, an evacuation shelter was established at a Buddhist temple for the purpose of sheltering local horses used for the annual festival.[57] To date (December 2014), only 30 percent of the town's evacuees have returned. However, the town has revived this celebration, one in which the mayor, with his samurai lineage, proudly participates.

Is there hope for genuine connectedness, or *tsunagari*? We see signs in younger mayors like Osaka's Hashimoto Tōru, who refused to restart local nuclear reactors in the face of strong pressure to do so, though he did eventually commission the restart of the Oi reactor in July 2012. And there is also Yubari's thirty-three-year-old mayor Suzuki Naomichi. His former coal-mining town in Hokkaido that now relies on nuclear and petroleum based energy, is bankrupt, crushed under a mountain of debt and a stagnating workforce. The town's population is heavily elderly. Yubari's tourism industry in its beautiful surroundings has nearly disappeared. Nonetheless, his energy, innovativeness, and youthful charisma inspires hope among his citizens. As

Hiroko Tabuchi writes in the April 26, 2012, *New York Times*, Yubari says that "In many ways, it's not my generation's fault that Japan has so much debt . . . But blaming others won't get us anywhere. We just need to find a way forward. *It's the responsibility of all of us born into this age*"[58] (author emphasis).

From my encounters with and working together with young volunteers, I believe that the youth who embody Japan's future give strong evidence of hope for the future. Not all feel the same. Notwithstanding the profoundly complex challenge in measuring well-being, the young writer Furuichi Noritoshi claims that as long as Japanese youth feel socially accepted, for example through Facebook friends, and other social media, and are recognized for wearing name-brand clothes, his generation feels "happy" enough despite any prevailing crisis. Furuichi is one of Japan's most promising young sociologists, and his recent best-selling book, *Zetsubōno Kuni no Kōfuku na Wakamono-tachi* (The Happy Youth of a Desperate Country), evoked considerable public response.[59] In it, he criticizes youth theories that categorize Japanese youth as either lazy or overburdened. Labeling them according to their age is unfair and simply ageist, overlooking factors such as gender, class, place of origin, and where they grow up. He argues that most young people (*wakamono*) throughout Japan are content with their lives. He admits that social connectivity generally disinclines youth to be more socially engaged. Even after 3.11, the extent to which *wakamono* commit themselves to social causes and volunteering is limited. True, many continue to volunteer, but what is more important is to lend support to their efforts through sound and effective policies.

To some degree, this makes sense. Yet it takes a strong dedicated core, though small in number, who can empower themselves in ways that makes the difference. This was certainly the case with the many volunteers I have met and worked with at Magokoro Net. They tell a different story. As a group, we have connected to help relieve others' suffering, whose suffering we have each somehow felt a need to experience. They share with me that shoveling, cleaning, and so forth, are only small tasks among "the dew of little things" they can do to help in rebuilding a fractured community. They affirm what Akiko repeatedly stressed—There is no recovery until everyone is recovered. By any measure, their sense of "happiness" runs deeper than having Facebook friends and wearing chic, stylish clothes. Genuine connectedness is possible, and it goes well beyond connectivity.

NOTES

1. This is no longer found. http://bouhatsusoshi.jp/english.
2. Skilled Veterans Corps for Fukushima, "Proposal for the participation of skilled veterans in the cleanup works of the Fukushima Daiichi Nuclear Power Plant accident," accessed Dec. 6, 2014, http://svcf.jp/english/proposal_en.
3. Cited by Lucy Craft, "Japanese Seniors, Send Us to Damaged Nuclear Plant," NPR (National Public Radio), accessed Dec. 4, 2014, http://www.npr.org/2011/09/12/140402430/japanese-seniors-send-us-to-damaged-nuclear-plant.
4. Emmanuel Levinas, "The philosopher and death," *Alterity & Transcendence*, ed. Pierre Hayat, trans. Michael B. Smith (New York: Columbia University Press, 1999), 161.
5. The Skilled Veterans Corps of Fukushima, "The fundamental status and capacity of Skilled Veterans' Corps for Fukushima, accessed Dec. 6, 2014, http://svcf.jp/english/fundamental_en.
6. Skilled Veterans Corps, "The fundamental status," accessed Dec. 6, 2014, http://svcf.jp/english/fundamental_en.
7. Roland Buerk, "Japan pensioners volunteer to tackle nuclear crisis," BBC News Asia-Pacific, May 31, 2011, accessed Dec. 5, 2014, http://www.dailygood.org/more.php?n=4629.
8. Buerk, "Japan pensioners volunteer," accessed. Dec. 5, 2014, http://www.dailygood.org/more.php?n=4629.
9. Lucy Birmingham and David McNeil, *Strong in the Rain: Surviving Japan's Earthquake, Tsunami, and Fukushima Nuclear Disaster* (New York: Palgrave Macmillan, 2012), 62.
10. It remains arguable whether the plant officials were considering completely abandoning the plant. In official hearings afterwards in Tokyo, there were mixed opinions.
11. Birmingham and McNeil, *Strong in the Rain*, 133.
12. Birmingham and McNeil, *Strong in the Rain*, 85.
13. Birmingham and McNeil, *Strong in the Rain*, 93. Linguistic scholars have also examined foreign media reporting. In her "The Fukushima Nuclear Crisis E-Coverage: A Linguistic Analysis of *Sciencemag.org* and *Scientificamerican.com*," Alessandra Vicentini offers an intriguing comparison of two online publications, the scientific *Sciencemag.org* and the semi-scientific *Scientificamerican.com* with respect to sentence structure, lexical choices, citation, and hedging. Her analysis combines critical discourse analysis and lexical analysis. Her chapter is in *The Popularization of Specialized Discourse and Knowledge Across Communities and Cultures*, ed. by Susan Kermas and Thomas Christiansen (Bari, Italy: Edipuglia, 2013): 297–314.
14. Birmingham and McNeil, *Strong in the Rain*, 94.
15. Birmingham and McNeil, *Strong in the Rain*, 97.
16. Michael C. Brannigan, "Just how safe is safe?" *Albany Times Union*, April 26, 2014, accessed January 28, 2015, http://www.timesunion.com/opinion/article/Michael-Brannigan-Just-how-safe-is-safe-5432505.php#page-1.
17. Gar Smith, *Nuclear Roulette: The Truth about the Most Dangerous Energy Source on Earth* (White River Junction, VT: Chelsea Green Publishing, 2012), 32.
18. Smith, *Nuclear Roulette*, 32.
19. World Nuclear Association, "Chernobyl Accident 1986," updated Dec. 2014, accessed Feb. 16, 2015, http://www.world-nuclear.org/info/Safety-and-Security/Safety-of-Plants/Chernobyl-Accident/.
20. U.S. NRC (Nuclear Regulatory Commission), "Backgrounder on Chernobyl Nuclear Power Plant Accident," May 2013, accessed Feb. 16, 2015, http://www.nrc.gov/reading-rm/doc-collections/fact-sheets/chernobyl-bg.html.
21. Alexy V. Yablokov, Vassily B. Nesterenko, and Alexy V. Nesterenko, "Chernobyl: Consequences of the Catastrophe for People and the Environment," *Annals of the New York Academy of Sciences* 1181 (December 2009), cited in Gar Smith, *Nuclear Roulette: The Truth about the Most Dangerous Energy Source on Earth* (White River Junction, VT: Chelsea Green Publishing, 2012), 241, note. 6.
22. World Nuclear Association, "Chernobyl Accident 1986," note b., http://www.world-nuclear.org/info/Safety-and-Security/Safety-of-Plants/Chernobyl-Accident/.

23. European Energy Forum, "In memory of Valery Legasov," Jan. 3, 2011, accessed Feb. 16, 2015, http://www.europeanenergyforum.eu/library/letter-from-publisher/memory-valery-legasov.

24. Birmingham and McNeil, *Strong in the Rain*, 82.

25. Birmingham and McNeil, *Strong in the Rain*, 189.

26. Anne Allison, *Precarious Japan* (Durham, NC: Duke University Press, 2013), 186.

27. Tom Gill, Brigitte Steger, and David H. Slater, "The 3.11 Disasters," in *Japan Copes with Calamity: Ethnographies of the Earthquake, Tsunami and Nuclear Disasters of March 2011*, 2nd ed., edited by Tom Gill, Brigitte Steger, and David H. Slater (Bern, Switzerland: Peter Lang, 2013, 2014), 15.

28. Gill *et al.*, "The 3.11 Disasters," 15ff.

29. Michael C. Brannigan, "How a culture of peace confronts violence," *Albany Times Union*, July 28, 2012, accessed January 28, 2015, http://www.timesunion.com/opinion/article/How-a-culture-of-peace-confronts-violence-3743527.php. This meant that his sentence will most likely extend beyond Norway's maximum 21 years if still considered a danger to society. Norway does not give the death penalty.

30. Birmingham and McNeil, *Strong in the Rain*, 199.

31. See Daniel P. Aldrich's eye-opening *Site Fights: Divisive Facilities and Civil Society in Japan and the West* (Ithaca, NY: Cornell University Press, 2008).

32. Jacques Le Goff, *The Birth of Purgatory*, trans. Arthur Goldhammer (Chicago: University of Chicago Press, 1984), 4; orig. *La nuisance du Purgatoire* (Paris: Editions Gallimard, 1981).

33. "Fukushima court rules against Tepco in suicide case," *Japan Times*, August 26, 2014, accessed Jan. 29, 2015, http://www.japantimes.co.jp/news/2014/08/26/national/fukushima-court-rules-tepco-blame-suicide-utility-ordered-pay-%C2%A549-million/#.VZ1fGV9Viko.

34. Yoko Wakatsuki and Madison Park, "TEPCO ordered to pay compensation in landmark Fukushima suicide case," *CNN*, August 27, 2014, accessed Jan. 28, 2015, http://www.cnn.com/2014/08/27/world/asia/japan-fukushima-health-effects/.

35. "Fukushima stress deaths top 3/11 toll," *Japan Times* (Feb. 20, 2014), accessed Jan. 15, 2015, http://www.japantimes.co.jp/news/2014/02/20/national/post-quake-illnesses-kill-more-in-fukushima-than-2011-disaster/#.VOPaMWctDcs.

36. Manami Amagai *et al.*, "Factors Related to Mental Health and Suicidal Thoughts of Adults Living in Shelters for a Protracted Period Following a Large-Scale Disaster," *Academic Journal of Interdisciplinary Studies* 3, No. 3 (June 2014): 11, accessed Jan. 15, 2015, http://www.mcser.org/journal/index.php/ajis/article/view/3378.

37. Birmingham and McNeil, *Strong in the Rain*, 177.

38. Interview with Minamisoma City Mayor Katsunobu Sakurai, Minamisoma, March 19, 2014.

39. Kawanishi Hidemichi, *Tōhoku o yomu* (Akita: Mumyōsha Shuppan, 2011): 20, in Nathan Hopson, "Systems of Irresponsibility and Japan's Internal Colony," *The Asia-Pacific Journal* 11, issue 52, no. 2 (December 30, 2013), accessed Feb. 6, 2015, http://japanfocus.org/-Nathan-Hopson/4053.

40. Birmingham and McNeil, *Strong in the Rain*, 63.

41. Daniel P. Aldrich, *Building Resilience: Social Capital in Post-Disaster Recovery* (Chicago: The University of Chicago Press, 2012).

42. Pierre Bourdieu, "The forms of capital," in *Handbook of Theory and Research for the Sociology of Education*, edited by John G. Richardson (New York: Greenwood Press, 1986), 241–258; Pierre Bourdieu, "Le capital social," *Actes de la Recherche en Sciences Sociales* 31 (1980): 2–3; Robert D. Putnam, *Bowling Alone: The Collapse and Revival of American Community* (New York: Simon & Schuster, 2001).

43. See eminent psychiatrist Robert J. Lifton's landmark and unforgettable, *Death in Life: The Survivors of Hiroshima* (Chapel Hill: University of North Carolina Press, 1991).

44. Gill *et al.*, "The 3.11 Disasters," 13–14.

45. Email message from Kunio Aoki, Tokyo, February 20, 2015.

46. Birmingham and McNeil, *Strong in the Rain*, 192.

47. Conversation with Kunio Aoki, Tochigi, March 20.

48. Birmingham and McNeil, *Strong in the Rain*, 171.
49. Birmingham and McNeil, *Strong in the Rain*, 174.
50. Daniel Aldrich and Yasuyuki Sawada, "The physical and social determinants of mortality in the 3.11 tsunami," *Social Science and Medicine*, 124 (January 2015): 66–75.
51. Interview with Makiko Okuyama, National Center for Child Health and Development, Tokyo, March 24, 2014. Aizu Wakamatsu remains a popular tourist attraction. It is the former castle town of the Aizu clan. There is also the gravesite of the legendary Byakkotai, 19 teenage samurai who committed ritual suicide upon seeing what they believed was the burning of the castle. This was part of the Aizu samurai's last stand leading to their defeat in the Boshin War. Their dead bodies were left to rot on the field as a warning to other rebellious clans. People come from all over Japan to this gravesite, drawn to the story of the young warriors. Ironically, the castle was not really burning. The smoke came from elsewhere.
52. "SOS from Mayor of Minami Soma City, next to the crippled Fukushima nuclear power plant, Japan," accessed Jan. 29, 2015, http://www.youtube.com/watch?v=70ZHQ--cK40.
53. Interview, Minamisoma City, March 19, 2014.
54. In addition to farming for twenty-six years, Sakurai is a strong sportsman. He feels that this helped him in facing the disaster. Since 3.11, he stayed in his office for fifty days. He released his stress by jogging, cold-sweating each morning followed by a shower. As he puts it, "Physical training is important to face your weakness."
55. Direct quotes are from the March 19, 2014, interview. All else is a paraphrase of our conversation and not a literal translation.
56. "HELP! SOS from Mayor Katsunobu Sakurai of Minami Soma City," accessed Jan. 10, 2015, https://www.youtube.com/watch?v=R6URqs9kb20.
57. Older horses are also a delicacy in Fukushima restaurants. My first and only time eating raw horse meat, called *sakura* due to its rich red color as in the cherry blossom, was at a Fukushima restaurant. *Oishii desu* (It was delicious).
58. Hiroko Tabuchi, "An Aging Japanese Town Bets on a Young Mayor for Its Revival," *New York Times*, April 26, 2012, accessed Jan. 10, 2015, http://www.nytimes.com/2012/04/27/business/global/aging-japanese-town-bets-on-a-young-mayor.html?pagewanted=all&_r=0.
59. Noritoshi Furuichi, *Zetsubōno Kuni no Kōfuku na Wakamono-tachi* (The Happy Youth of a Desperate Country) (Tokyo: Kodansha, 2011).

Conclusion

The Big One

Brooke and I loved visiting Portland, Oregon. There, life wears a different set of clothes. It takes on its own unique tempo. Portland is a fairly-sized city with its traffic, highways, bustle, and business. But, it's the people. Civil, even friendly, always with a greeting, a bit laid back, taking life in stride—not the norm here in our part of the Northeast. We especially loved the coastline, about an hour away. Cannon Beach stunned us with its raw beauty, an enclave of sandy beach, a small, quaint village, and prominently pointed and powerful rocks jutting out close to shore as if they were catching the waves. Views from the steep cliffs were mesmerizing. We loved it here. Would this turn out to be Our Place, that lingering question couples carry when traveling beyond their familiar? Reality checks always cash in, however, usually in the form of finances and economy. Another is the sign posted on the beach—"In case of tsunami . . . "

Residents of Oregon's coastline communities are familiar with these customary warnings, what Japanese call *Abunai!* And for good reason. Experts, seismologists, climatologists and their ilk warn of the imminent Big One, sometime early this century. It's the same Big One that could slice off a huge piece of the West Coast, from Washington to Southern California. California has its San Andreas Fault, and the Pacific Northwest coast, sitting parallel to the Cascadia subduction zone, has a fault that runs six hundred miles along that coastline up to British Columbia.[1] It's the Big One that swallowed a good piece of Sri Lanka and Indonesia on December 26, 2004, as hundreds of thousands met their watery graves in the Indian Ocean. Over 230,000 perished. It's the One that battered New Orleans, a.k.a Katrina, towards the end of August 2005 that drowned portions of Louisiana, Mississippi, and

Alabama. Over 1,800 died. In the United States, Katrina remains our benchmark for how we respond and not respond. The Big One wiped out over 68,000 people in China's Sichuan Province's May 12, 2008, earthquake. It later demolished Haiti on January 12, 2010, with an earthquake killing well over 200,000 people.

For communities who live by the sea, which is always predictably unpredictable and can shift moods in seconds for better or for worse, they face sure risks. Fishermen and sailors know this all too well. And they are reminded of this when they suddenly find themselves trapped in a wet veil of fog, usually on their way back to land. I've learned this the hard way kayaking. As I paddle out towards the sea from nestled Bissel Cove near Wickford, Rhode Island, the apparent calm can easily trick you into thinking it will stay that way. When it suddenly changes, the trip back under dark, sinister clouds is daunting, and when powerful currents lift the stern higher than you'd like, you either enjoy the ride or paddle back white-knuckled as quickly as you can.

For Japanese in Tohoku who have chosen to be the sea's neighbor, the specter of the Big One always lingers even if slightly in the back of their minds, perhaps far back in consciousness. Indeed, this is especially so in Japan, far more than in other countries. Sitting in the midst of the Pacific Ring of Fire, home to over one hunred active volcanoes and spewing around 1500 tremors a year, Japan's history, past and recent, stubbornly reminds its people that the next Big One is never too far away. Each year, on September 1, Japan officially celebrates National Disaster Prevention Day. On this day, in 1923, a 7.8 magnitude earthquake produced both a tsunami with waves as high as forty feet and a terrible conflagration that spread throughout the region, all combining to level Tokyo and Yokohama, killing over one hundred thousand people. My mother, Misae Kimura, escaped this by five years and over five hundred miles. She was born and grew up in Fukuoka, on the southern island of Kyushu, in July 1928. Yet, as with all of us, there were other threats she could not sidestep.

The Pacific War was intensely onerous with the smell of mass death. The incendiary bombing of Tokyo caused over 100,000 deaths, up until then the war's largest mass killing. That is, until August 1945's Hiroshima and Nagasaki, which together incinerated nearly 500,000 victims, nearly all civilians. Mom would tell us stories of kneeling under her school desk when sirens screeched during air-raid drills. Sometimes classes would evacuate the school building under teachers' watchful eyes. These drills came in handy years later when American bombers flew over her hometown on their way to torch Tokyo. At the scream of the sirens, her family ran to their small backyard, opened the hatchdoor in the ground, and crammed themselves into their tiny homemade, hand-dug shelter, a hole in the ground just big enough to fit her parents, brother, three sisters, and herself. There they stayed, squeezed together, frozen in fear until the planes left, wondering if their own Big One would arrive uninvited at any moment.

We humans are strange. We can live with incessant reminders and still disregard the probable, going about our business-as-usual. In a sense, we need to. We cannot live and function in ever-present fear and anxiety over what could occur years, months, weeks, days, or hours from now, tonight, or tomorrow. This sort of collective avoidance is no doubt global, and is certainly the case here in the United States. Yet complete disregard is hazardous to one's health. We cannot afford to ignore lessons of the past. Ignorance and denial worsens the "crime," whether an act of God or act of man. Though faces and names change, victims remain the same. Our disregard, ignorance, or denial is what Jane Jacobs calls the Fifth Horseman of the Apocalypse—"Forgetfulness." Under the tutelage of Forgetfulness we bring the Dark Ages upon us, and this simply perpetuates the crime. Here, again, is where moral grit enters. It is not about living in some constant state of apprehension, fixed to what *could* be. Moral grit demands constructive defiance in the face of darkness so that we commit ourselves entirely to the present to secure a better future. Albert Camus says it well when he states that "Our best investment in the future is to give entirely to the present."

Moral grit uncovers our most telling lesson. Namely, who we are and how we are, personally and with each other, *before* the storm plays out, *during*, and *after* the storm. If business-as-usual means back to each-for-himself, then we've woefully never learned our lesson. My mother's retreat to her backyard "shelter" was her family's retreat, not hers alone. They each faced life, affliction, and the threat of extinction not alone, but together.

WHAT'S IN STORE: THE EARTHQUAKE OF 201X AND EXTREME WEATHER

According to the researchers at the University of Tokyo's Earthquake Research Institute, there is a 70 percent likelihood of a huge earthquake occurring close to Tokyo within a decade, perhaps as early as 2016. A continuous series of smaller quakes in the aftermath of 3.11 portend a massive quake to come. These smaller shakes are the constant hiccups, a "mathematical omen," leading to one huge heave.[2] Ongoing friction between the Philippine Sea Plate and the Eurasia Plate pretty much foretells a monumental rupture in the not too-distant future. The Big One is expected in the form of a magnitude 7 earthquake labelled the Tokai Earthquake. This is expected in an area where earthquakes have regularly struck every one hundred to 150 years in what is Japan's most volatile area for earthquakes, on the Pacific coastline of Honshu just southwest of Tokyo, in Shizuoka Prefecture, in the area of

Suruga Bay, called the Tokai segment. This Tokai segment is part of the subduction zone where the Philippine Sea Plate rubs under the Eurasia Plate. This is perhaps the most closely monitored region in Japan for earthquakes and has an historical pattern with enormous earthquakes occurring in 1498, 1605, 1707, and 1854. With the last in 1854, the next Big One is long overdue.

Renowned seismologist Katsuhiko Ishibashi first brought this warning of the Big One to professional and public attention in 1976. Since then, the government has stepped up measures to enhance earthquake response education and to alleviate future earthquake damage. And no doubt since 3.11, there have been more exacting safety precautions to anticipate the worst case scenario for Tokyo, in which projected thirty meter (over ninety feet) waves could bring about over 300,000 deaths. Imagine the waters flooding Tokyo's labyrinthine underground subway stations, a complete shutdown of power, and fires from cracked fuel tanks ravaging the huge metropolis.[3] Hitting Tokyo will produce full-scale catastrophe. And, according to researchers, this is imminent. Moreover, the Chubu Electric Power Company's Hamaoko nuclear power plant sits in this volatile area, situated in Omaezaki, Shizuoka Prefecture. The plant's buildings have been reinforced to become more earthquake resistant. It continues to operate despite growing public opposition. Nonetheless, Tokyo is going full-steam ahead preparing for the 2020 Summer Olympics.

Even worse, a Tokai earthquake may occur in combination with what are called "Tonankai" and "Nankai" earthquakes, named after regions close to Tokai. This would be truly catastrophic, tripling the fatalities from Shizuoka to Kochi prefectures to easily over 350,000.[4] According to disaster risk reduction expert Yoshiaki Kawata, "The prediction is more severe than the death toll from the Great East Japan Earthquake. But the 400,000 figure is not groundless conjecture. I want people to fully understand how dangerous [such a disaster] would be and to think about disaster prevention measures."[5]

Experts may disagree on what to expect in the near future. However, one thing is certain regarding weather and natural disasters. There will always be a next time. And it will be worse.[6]

In the summer of 2003, Europe experienced its worst natural disaster on record. Extreme heat scorched major portions of Europe, killing over 35,000 people. In France alone, nearly 15,000 succumbed to the unbearable heat. High temperatures set off forest fires in Portugal. And melting ice in the Alps spawned flash floods and huge avalanches.[7] Climate experts claim that this was indeed a freak happening, beyond expectations and forecasts. Nonetheless, they also predict that such anomalies will become more commonplace in the decades ahead, not centuries from now, but much sooner. Climatologist Heidi Cullen is straightforward in laying it out:

> By 2050, mid-range emissions scenarios predict that a day so hot that it is currently experienced only once every twenty years would occur every three years over much of the continental United States. By the end of the century, such a day would occur every other year, or more often. As for the cold, we'll be seeing even less of it. By 2100, the number of frost days averaged across North America is projected to decrease by one month; and decreases of more than two months are projected in some places.[8]

Cullen does not stop there. She goes on to point out that warmer temperatures will also increase evaporation so that our atmosphere will contain more moisture. Areas in the United States, notably the West and Southwest, will suffer increased droughts, and the Northeast, Midwest, and Northwest will encounter heavier rain, blizzards, and flooding. Warmer weather and increased evaporation also means warmer oceans and more hurricanes, hurricanes that will be much more intense. And with more people moving to coastlines experiencing rising sea levels, we have the recipe for ongoing disaster. A "crime" waiting to happen.

It is clear what we must do. Again, we can point to global warming as the culprit. But many of us, to our peril, will overlook human causation. Unless we curb the rise in global population and increasing use of fossil fuels like natural gas, oil, and charcoal, all combining to build up greenhouse gases in the atmosphere and deplete our ozone layer, we will inevitably bring about our own demise. T. S. Eliot warns us in his immortal conclusion to "The Hollow Men," "This is the way the world ends, Not with a bang but a whimper." We are undeniably at a crossroads, and we cannot afford to wait until the waves are just hours away, when the train is about to cross our path. There you have it. The United States will become warmer, less frigid, wetter, and dryer. Yet, as far as we are individually concerned, this is far enough from our own personal futures. We can still buy that ocean front cottage in Florida. And many of us will move south to escape the cold.

UNDYING HOPE

What we need to do is obvious. But equally vital, we must seriously ask ourselves, "Can we cultivate genuine community, connectedness, before the storm"? Nature has a way of slapping us hard in the face. But we never listen. Finally, she lets loose the Big One and flips our world upside down. Disasters remind us that we are not the measure of all things. That title goes to the unbending rhythm of life, its cycle of birth, death, and all that lies in-between. Yet when the well is dried, what lies within us carries the day. Our inner spirit and stamina is the measure of our moral grit. Disaster is the litmus test, the ultimate test that challenges our personal and social fabric. The neighboring communities of Fargo, North Dakota, and Moorhead, Minnesota chose to act in tandem to mitigate forecasted flooding from the Red River that was expected to crest over twenty-four feet above flood level. They worked together side-by-side as a single community. They did not wait for the disaster to come to them. 350,000 cubic yards of dirt and 3.5 million sandbags later, they managed to stave off what would have been a colossal nightmare.[9]

With respect to Japan, while it had become increasingly atomized through consumerism and its private competitive ethos, and in turn more socially lethargic, the 3.11 crisis may have shocked this "lonely crowd" out of its lethargy.[10] Cultural critic and professor at Waseda University in Tokyo, Hiroki Azuma, views the disaster as a turning point, inciting hope, *kibō*, and pride in a nation that had slowly lost it. He also believes that this hope is especially apparent among the younger generation. As he put it soon after the triple impact, "To borrow an expression from the younger generation here, the Japanese people seem to have completely transformed their *kyara* (character)."[11]

In past decades, there appeared to be little reason for hope. Yet, 3.11 forced a turnabout. Perhaps this is less of a "turn" about but more of a "return" together in solidarity, in a spirit of belonging. To "return" means to go back to where one came from. In conversation, Japanese do not say "to *go* home," but "to *return* home," *uchi ni kaerimasu*. Home is the original source of belonging. The many with whom I met strongly affirm this. The events of 3.11 have forced their hand at cultivating hope and moral grit, steering them to re-embrace their source, their home, and Nature. Survivors, city leaders, volunteers, and others with whom I met in Japan have imparted to me this unbreakable resilience of the human spirit. Many of them have taught me lessons in moral grit. Survivors especially have shown me that without moral grit, like many of them, we face our condition in isolation, and we turn generic.

In his later years, the great artist Francisco José de Goya y Lucientes (1746–1828), whose paintings often reflect the dark side of life, suffered debilitating illness. Yet despite his chronic pain and weakness, poor eyesight and hearing, failing health, and family deaths, he once told a friend, "Neither sight, nor pen nor inkwell; all these I lack, and all that is plentiful is my will."[12] Here is the grit he harbors in the face of his condition.

It is now impossible for me to return to our version of what's real. Though I'm back in our safe rendering of the real, I'm no longer the same. For me, split apart between two island cultures, Japan and Ireland, I feel all the more country-less. After shoveling snow and mud in Tohoku, there is no going back to a place where our prevailing version of reality—living in our screens, consumed by the power of the Image, infatuated with celebrities, money, status, and falling deeper into the rabbit hole of the hyperreal—has become even more saccharin, artificial.

Amidst our lives saturated with uncertainties, however, one thing remains certain. Each shovelful is merely a footstep that helps to lift another's burden and eases another's anxiety. Each shovelful makes up the wheelbarrow, and each wheelbarrow makes up the mountain. Each shovelful makes all the difference and together, brings us closer to home and to what matters.

NOTES

1. Tim Folger, "The Calm before the Wave," *National Geographic* 221, issue 2 (February 2012): 71.
2. Yoree Koh, "Big Tokyo Quake Forecast by 2016," *The Wall Street Journal*, Jan. 23, 2012, accessed Feb. 2, 2015, http://blogs.wsj.com/japanrealtime/2012/01/23/big-tokyo-quake-forecast-by-2016/.
3. Lucy Birmingham, "Two Years after Fukushima, Japan Worries about the Next Big Quake," *Time*, March 10, 2013, accessed Feb. 5, 2015, http://world.time.com/2013/03/10/two-years-after-fukushima-japan-worries-about-the-next-big-quake/.
4. "New prediction shows triple quake to kill 400,000 in Japan," *Yomiuri Shimbun*, July 11, 2012, accessed Feb. 11, 2015, http://www.asianewsnet.net/news-33210.html. See also Peter Hadfield, *The Coming Tokyo Earthquake: Sixty Seconds That Will Change the World* (Tokyo: Tuttle, 1995).
5. "New prediction," *Yomiuri Shimbun*, http://www.asianewsnet.net/news-33210.html.
6. T. R. Karl et al., *Weather and Climate Extremes in a Changing Climate: Regions of Focus* (U.S. Climate Change Science Program, 2008), cited in Heidi Cullen, *The Weather of the Future: Heat Waves, Extreme Storms, and Other Scenes from a Climate-Changed Planet* (New York: Harper, HarperCollins Publishers, 2010), 304, note 3.
7. Cullen, *Weather of the Future*, 51.
8. Cullen, *Weather of the Future*, 51.
9. Cullen, *Weather of the Future*, 3–4.
10. See David Riesman's classic sociological study of contemporary American society in its revised edition, David Riesman, Nathan Glazer, and Reuel Denney, *The Lonely Crowd*, rev. ed. (New Haven, CT: Yale University Press, 2001); orig. publication 1961.
11. Hiroki Azuma, "For a Change, Proud to Be Japanese," *New York Times* (March 16, 2011): A35.
12. J. M. Matilla, *Goya* (Madrid, Spain: TF Arts Gráficas, 2000), cited in Polyxeni Potter, "Francisco José de Goya y Lucientes (1746–1828). Self-portrait with Dr. Arrieta (1820)," *Emerging Infectious Diseases* 10, no. 5 (May 2004): 974.

Bibliography

Aldrich, Daniel P. *Site Fights: Divisive Facilities and Civil Society in Japan and the West*. Ithaca, NY: Cornell University Press, 2008.

———. *Building Resilience: Social Capital in Post-Disaster Recovery*. Chicago, IL: University of Chicago Press, 2012.

Aldrich, Daniel, and Yasuyuki Sawada. "The physical and social determinants of mortality in the 3.11 tsunami." *Social Science and Medicine* 124 (2015): 66–75.

Allison, Anne. *Precarious Japan*. Durham, NC: Duke University Press, 2013.

Amagai, Manami, Noriko Kobayashi, Mayumi Nitta, Makiko Takahashi, Ikuko Takada, Yumiko Takeuch, Yumiko Sawada, and Mayo Hiroshima. "Factors Related to Mental Health and Suicidal Thoughts of Adults Living in Shelters for a Protracted Period Following a Large-Scale Disaster." *Academic Journal of Interdisciplinary Studies* 3 (2014): 11–16. At http://www.mcser.org/journal/index.php/ajis/article/view/3378.

Aoki, Shinmon. *The Journal of a Buddhist Mortician: Coffinman* [*Nokanfu nikki*]. Trans. Wayne S. Yokoyama. Anaheim, CA: Buddhist Education Center, 2002.

Arima, Tatsuo. *The Failure of Freedom: A Portrait of Modern Japanese Intellectuals*. Cambridge, MA: Harvard University Press, 1969.

Ashcroft, Frances. *Life at the Extremes*. London: Flamingo, HarperCollinsPublishers, 2001.

Averbuch, Irit. "Modern Meanings of Kagura Performances." *Dharma World* (2013). At http://www.rk-world.org/dharmaworld/dw_2013aprjune_modern-meanings-of-kagura-performances.aspx. Also at http://www.xip.fi/atd/japan/kagura-sacred-shinto-dances.html.

Azuma, Hiroki. "For a Change, Proud to Be Japanese." *New York Times* (March 16, 2011): A35.

Barbisch, Donna F., and Kristi Koenig. "Understanding Surge Capacity: Essential Elements." *Academic Emergency Medicine* 13 (2006): 1098–1102.

Bassino, Jean-Pascal. "Market Integration and Famines in Early citizens Modern Japan, 1717–1857." 1–22. At http://federation.ens.fr/ydepot/semin/texte0708/BAS2007MAR.pdf.

Baudrillard, Jean. *Simulations*. New York: Semiotext(e), 1983.

Bauman, Zygmunt. *Culture in a Liquid Modern World*. Trans. (from Polish) by Lydia Bauman. Cambridge, UK: Polity Press, 2011.

Bein, Steve, trans. and commentary. *Purifying Zen: Watsuji Tetsurō's Shamon Dōgen*. Honolulu, HI: University of Hawaii Press, 2011.

Birmingham, Lucy. "Two Years after Fukushima, Japan Worries about the Next Big Quake." *Time* (March 10, 2013). At http://world.time.com/2013/03/10/two-years-after-fukushima-japan-worries-about-the-next-big-quake/.

Birmingham, Lucy, and David McNeil. *Strong in the Rain: Surviving Japan's Earthquake, Tsunami, and Fukushima Nuclear Disaster*. New York: Palgrave Macmillan, 2012.

Blaikie, Piers, Terry Cannon, Ian Davis, and Ben Wisner. *At Risk: Natural Hazards, People's Vulnerability, and Disasters*, 2nd ed. London: Routledge, 2003.

Bourdieu, Pierre. "Le capital social." *Actes de la Recherche en Sciences Sociales* 31 (1980): 2–3.

———. "The forms of capital." 241–58 in *Handbook of Theory and Research for the Sociology of Education*. Edited by John G. Richardson. New York: Greenwood Press, 1986.

Brannigan, Michael C. "A Chronicle of Organ Transplant Progress in Japan." *Transplant International* 5 (1992): 180–86.

———. *The Pulse of Wisdom: The Philosophies of India, China, and Japan*, 2nd ed. Belmont, CA: Wadsworth Publishing Company, Inc., 1995.

———.*Cultural Fault Lines in Health Care: Reflections on Cultural Competency*. Lanham, MD: Lexington Books, Rowman & Littlefield Publishers, Inc., 2012.

———."Our Nightingales' Song from Tohoku." *Japan Railway & Transport Review* 60 (2012): 60–62.

———."How a culture of peace confronts violence." *Albany Times Union* (July 28, 2012). At http://www.timesunion.com/opinion/article/How-a-culture-of-peace-confronts-violence-3743527.php.

———."Just how safe is safe?" *Albany Times Union* (April 26, 2014). At http://www.timesunion.com/opinion/article/Michael-Brannigan-Just-how-safe-is-safe-5432505.php#page-1

———."Let torch illuminate solidarity." *Albany Times Union* (June 26, 2014). At http://www.timesunion.com/opinion/article/Michael-Brannigan-Let-torch-illuminate-solidarity-5582346.php

———."Guarding against a deadly outbreak of unfounded fear." *Albany Times Union* (Nov. 26, 2014). At http://www.timesunion.com/opinion/article/Michael-Brannigan-Guarding-against-a-deadly-5920030.php.

Brannigan, Michael C. (Kenji Kimura). "One Shovelful at a Time—Volunteering in Tohoku." *Japan Railway & Transport Review* 65 (2015): 44–47.

Buerk, Roland. "Japan pensioners volunteer to tackle nuclear crisis." BBC News Asia- Pacific (2011). At http://www.dailygood.org/more.php?n=4629.

Britton, Neil R. "Dog or Demon." 113—21 in *What Is a Disaster? New Answers to Old Questions*. Edited by Ronald W. Perry and E. L. Quarantelli. International Research Committee on Disasters: Xlibris Corporation, 2005.

Carter, Robert E. *Encounter with Enlightenment: A Study of Japanese Ethics*. Albany, NY: State University of New York Press, 2001.

Clancey, Gregory. *Earthquake Nation: The Cultural Politics of Japanese Seismicity, 1868—1930*. Berkeley: University of California Press, 2006.

Cohen, Stanley. *States of Denial: Knowing about Atrocities and Suffering*. Cambridge, UK: Polity Press, Blackwell Publishing, 2001.

Coontz, Robert. "Comparing Earthquakes, Explained." At http://news.sciencemag.org/2011/03/comparing-earthquakes-explained.

Craft, Lucy. "Japanese Seniors, Send Us to Damaged Nuclear Plant." National Public Radio (2014). At http://www.npr.org/2011/09/12/140402430/japanese-seniors-send-us-to-damaged-nuclear-plant.

Cullen, Heidi. *The Weather of the Future: Heat Waves, Extreme Storms, and Other Scenes from a Climate-Changed Planet*. New York: Harper Collins Publishers, 2010.

Deraniyagala, Sonali. *Wave*. New York: Vintage Books, Random House, LLC, 2013.

Dower, John. *Embracing Defeat: Japan in the Wake of World War II*. New York: W. W. Norton & Company, 1999.

Erhlich, Gretel. *Facing the Wave: A Journey in the Wake of the Tsunami*. New York: Pantheon Books, 2013.

European Energy Forum. "In memory of Valery Legasov." Jan. 2011. At http://www.europeanenergyforum.eu/library/letter-from-publisher/memory-valery-legasov.

Fan, Peggy. "Building a Resilient Volunteer Organization." *Stanford Social Innovation Review* (2014). At http://www.ssireview.org/blog/entry/building_a_resilient_volunteer_organization.

Folger, Tim. "The Calm before the Wave." *National Geographic* 221 (2012): 54—77.

Fossum, Anne Cecilie. Interview. Oslo, Norway. June 21, 2012.

Fox, Renée C. *Doctors without Borders: Humanitarian Quests, Impossible Dreams of Médecins Sans Frontières*. Baltimore, MD: Johns Hopkins University Press, 2014.

Frankenberg, Elizabeth, Thomas Gillespie, Samuel Preston, Bondan Sikoki, and Duncan Thomas. "Mortality, the Family and the Indian Ocean Tsunami." At http://eml.berkeley.edu/~webfac/alexroth/dthomas.pdf.

"Fukushima court rules against Tepco in suicide case." *Japan Times*. August 26, 2014. At http://www.japantimes.co.jp/news/2014/08/26/national/fukushima-court-rules-tepco-blame-suicide-utility-ordered-pay-%C2%A549-million/#.VZ1fGV9Viko.

"Fukushima stress deaths top 3/11 toll." *Japan Times*. February 20, 2014. At http://www.japantimes.co.jp/news/2014/02/20/national/post-quake-illnesses-kill-more-in-fukushima-than-2011-disaster/#.VOPaMWctDcs.

Furuichi, Noritoshi. *Zetsubōno Kuni no Kōfuku na Wakamono-tachi* (The Happy Youth of a Desperate Country). Tokyo: Kodansha, 2011.

Gibran, Kahlil. *The Prophet*. New York: Alfred A. Knopf, Inc., 1923.

Giddens, Anthony. *Runaway World: How Globalisation is Reshaping Our Lives*. New York: Routledge, 2003.

Gilhooly, Robert. "Suicides upping casualties from Tohoku disaster." *Japan Times*. June 23, 2011. At http://www.japantimes.co.jp/news/2011/06/23/national/suicides-upping-casualties-from-tohoku-catastrophe/#.VJSnq8BEA.

———."Fukushima man opts to be guinea pig." *Japan Times*. Sept. 14, 2011. At http://www.japantimes.co.jp/news/2011/09/14/national/fukushima-man-opts-to-be-guinea-pig/#.VOE0OWctDcs.

Gill, Tom, Brigitte Steger, and David H. Slater, eds. *Japan Copes with Calamity: Ethnographies of the Earthquake, Tsunami and Nuclear Disasters of March 2011*, 2nd ed. Bern, Switzerland: Peter Lang, 2014.

Gill, Tom, Brigitte Steger, and David H. Slater. "The 3.11 Disasters." 3–23 in *Japan Copes with Calamity: Ethnographies of the Earthquake, Tsunami and Nuclear Disasters of March 2011*. 2nd edition. Edited by Tom Gill, Brigitte Steger, and David H. Slater. Bern, Switzerland: Peter Lang, 2014.

Gjørv, Alexandra Bech. Interview. Oslo, Norway. May 12, 2014.

Gulati, Ranjay, Charles Casto, and Charlotte Krontiris. "How the Other Fukushima Plant Survived." *Harvard Business Review* (2014): 111–15. At https://hbr.org/2014/07/how-the-other-fukushima-plant-survived.

Hadfield, Peter. *The Coming Tokyo Earthquake: Sixty Seconds That Will Change the World*. Tokyo: Tuttle, 1995.

Hara, Katsurō. *An Introduction to the History of Japan*. New York: G.P. Putnam's Sons, 1920.

Hesse, Hermann. *The Glass Bead Game* (*Magister Ludi*)> Trans. Richard and Clara Winston. New York: Owl Books, Henry Holt, 1990.

Hewitt, K. *Regions of Risk. A Geographical Introduction to Disasters*. London:Addison, Wesley Longman, 1997.

Hidemichi, Kawanishi. *Tōhoku o yomu*. Akita: Mumyōsha Shuppan, 2011.

Hirose, Takashi. *Fukushima Meltdown: The World's First Earthquake-Tsunami-Nuclear Disaster*. Trans. Izumi Kaori, Alice Kawauchi, C. Dounglas Lummis, Matt Kazuko, Niwa Shinobu, Okuda Hiroyuki, Ono M. Yoko, Otsuka Teruyo, and Yoko. Orig. *Fukushima Genpatsu Merutodaun*. Tokyo: Asahi Shimbun Publications, Inc., 2011.

Hoffman, Susanna M., and Anthony Oliver-Smith, eds. *Catastrophe and Culture: The Anthropology of Disaster*. Santa Fe, New Mexico: School of American Research Press, 2002.

Holen, Are. Interview. Oslo, Norway. June 20, 2012.

Honda, Toshiaki. Interview. Tono, Japan. March 5, 2014.

Hopson, Nathan. "Systems of Irresponsibility and Japan's Internal Colony." *The Asia-Pacific Journal* 11 (2013). At http://japanfocus.org/-Nathan-Hopson/4053.

Independent Investigation Commission on the Fukushima Nuclear Accident. *The Fukushima Daiichi Nuclear Power Station Disaster: Investigating the Myth and Reality*. Edited by Mindy Kay Bricker. London: Routledge, 2014.

"Japan Dotted with Volcanoes that Could Erupt at Any Time." *Nikkei Asian Review*. At http://asia.nikkei.com/Tech-Science/Science/Japan-dotted-with-volcanoes-that-could-erupt-at-any-time.

"Japan–366 Days after the Quake . . . 19,000 lives lost, 1.2 million buildings damaged, $574 billion." Earthquake-Report.Com. At http://earthquake-report.com/2012/03/10/japan-366-days-after-the-quake-19000-lives-lost-1-2-million-buildings-damaged-574-billion/.

Joint Commission on Accreditation of Healthcare Organizations. *Healthcare at the Crossroads: Strategies for Creating and Sustaining Community-wide Emergency Preparedness Strategies*. Oakbrook Terrace, IL: Joint Commission on Accreditation of Healthcare Organizations, 2003.

Jones, Carolyn A., and Peter Galison, eds. *Picturing Science, Producing Art*. London: Routledge, 1998.

Kamiya, Setsuko. "Students Credit Survival to Disaster-Preparedness Drills." *Japan Times*. June 4, 2011. At http://www.japantimes.co.jp/news/2011/06/04/national/students-credit-survival-to-disaster-preparedness-drills/#.VMeuQmctDcs.

Karl, T. R., et al. *Weather and Climate Extremes in a Changing Climate: Regions of Focus*. U.S. Climate Change Science Program. 2008.

Kazuhiro, Soda. "*Seishin* (Mental)." *Berlinale* (2009): 140–43. At https://www.berlinale.de/external/de/filmarchiv/doku_pdf/20091068.pdf.

Kelly, William W. "Tohoku's Future: Predicting Outcomes or Imagining Possibilities?" *The Asian-Pacific Journal: Japan Focus*. At http://japanfocus.org/-William_W_-Kelly/3703.

Kingston, Jeff, ed. *Natural Disaster and Nuclear Crisis in Japan: Response and Recovery after Japan's 3/11*. London: Routledge, 2012.

Klinenberg, Eric. *Heat Wave: A Social Autopsy of Disaster in Chicago*. Chicago, IL: University of Chicago Press, 2002.

Klinenberg, Eric. Interview. At http://press.uchicago.edu/Misc/Chicago/443213in.html.

Koh, Yoree. "Big Tokyo Quake Forecast by 2016." *The Wall Street Journal* (Jan. 23, 2012). At http://blogs.wsj.com/japanrealtime/2012/01/23/big-tokyo-quake-forecast-by-2016/.

Kohut, A., S. Keeter, C. Doherty, M. Dimock, and M. Remez. "Economy, Jobs Trump All Other Policy Priorities." Washington, DC: Pew Research Center for the People and the Press, 2009.

Latour, Bruno. "How to Be Iconophilic in Art, Science, and Religion?" 418–40 in *Picturing Science, Producing Art*. Edited by Carolyn A. Jones and Peter Galison. London: Routledge, 1998.

Lear, Jonathan. *Radical Hope: Ethics in the Face of Cultural Devastation*. Cambridge, MA: Harvard University Press, 2006.

Le Goff, Jacques. *The Birth of Purgatory*. Trans. Arthur Goldhammer. Chicago: University of Chicago Press, 1984. Orig. *La nuisance du Purgatoire*. Paris. Editions Gallimard, 1981.

Levinas, Emmanuel. "The philosopher and death." 153–68. *Alterity & Transcendence*. Edited by Pierre Hayat, trans. Michael B. Smith. New York: Columbia University Press, 1999.

———.*Humanism of the Other*. Trans. Nidra Poller. Urbana: University of Illinois Press, 2006.

Lifton, Robert J. *Death in Life: The Survivors of Hiroshima*. Chapel Hill, NC: University of North Carolina Press, 1991.

Lochbaum, David, Edwin Lyman, Susan Q. Stranahan, and the Union of Concerned Scientists. *Fukushima: The Story of a Nuclear Disaster*. New York: The New Press, 2014.

Luke, Elmer, and David Karashima. *March Was Made of Yarn: Reflections on the Japanese Earthquake, Tsunami, and Nuclear Meltdown*. New York: Vintage, Random House, Inc., 2012.

MacFarlane, Alan. "The three major famines of Japanese history." At http://www.alanmacfarlane.com/savage/A-JAPFAM.PDF.

Matilla, J. M. *Goya*. Madrid, Spain: TF Arts Gráficas, 2000.

Mauch, Christof, and Christian Pfister, eds. *Natural Disasters, Cultural Responses: Case Studies toward a Global Environmental History*. Lanham, MD: Lexington Books, Rowman & Littlefield Publishers, Inc., 2009.

Miura, Hideyuki. "Healthy Forests Key to Rebuilding Tohoku Communities." *Asahi Shimbun* (April 3, 2012).

Miyazawa, Kenji. *Strong in the Rain: Selected Poems*. Trans. Roger Pulvers. Northumberland, UK: Bloodaxe Books, 2007.

———. "Translating Japanese: Strong in the Rain." Trans. Roger Pulvers. At http://mooninthewater.net/aikido/2012/10/13/translating-japanese-strong-in-the-rain/.

Murakami, Ryū. *Kibō no kuni no ekosodasu* [The Exodus of a Country with Hope]. Tokyo: Bunshun Bunko, 2002.

National Catalogue of the Active Volcanoes in Japan, July 2013. At http://www.data.jma.go.jp/svd/vois/data/tokyo/STOCK/souran_eng/menu.htm and https://gbank.gsj.jp/volcano-QV/VOL_JP/EN/active_v.htm.

National Police Agency of Japan. "Damage Situation and Police Countermeasures." At http://www.npa.go.jp/archive/keibi/biki/higaijokyo_e.pdf.

Nishimura, Akira. "Battlefield Pilgrimage and Performative Memory: Contained Souls of Soldiers in Sites, Ashes, and Buddha Statues." *Memory Connection* 1 (2011): 302–11. At http://memoryconnection.org/wp-content/uploads/2011/12/AkiraNishimura1.pdf.

Nordstrom, Carolyn. *Global Outlaws: Crime, Money, and Power in the Contemporary World.* Berkeley: University of California Press, 2007.

Norges offentlige utredninger. *Rapport fra 22. Juli-kommisjonen.* Oslo: Departementenes servicessenter Informasjonforvaltning, 2012. [Norwegian Public Reports. *Report from The 22. July Commission.* Oslo: Government Service Center Information Management, 2012.]

Odin, Steve. *The Social Self in Zen and American Pragmatism.* Albany: State University of New York Press, 1996.

Oe, Kenzaburo. *A Personal Matter*. Trans. John Nathan. New York: Grove Press, Inc., 1969.

Okuyama, Makiko. Interview. National Center for Child Health, Tokyo. March 24, 2014.

Oliver-Smith, Anthony, and Susanna M. Hoffman, eds. *The Angry Earth: Disaster in Anthropological Perspective.* New York: Routledge, 1999.

———. "Why Anthropologists Should Study Disasters." 3-22 in *Catastrophe and Culture: The Anthropology of Disaster*. Edited by Susanna M. Hoffman and Anthony Oliver-Smith. Santa Fe, New Mexico: School of American Research Press, 2002.

Onishi, Norimitsu. "Seawalls Offered Little Protection Against Tsunami's Crushing Waves." *New York Times* (March 13, 2011). At http://www.nytimes.com/2011/03/14/world/asia/14seawalls.html?pagewanted=all&_r=0.

Orui, Masatsugu, Shuichiro Harada, and Mizuho Hayashi. "Changes in suicide rates in disaster-stricken areas following the Great East Japan Earthquake and their effect on economic factors: an ecological study." *Environmental Health and Preventive Medicine* 19 (2014): 459-66.

Nakamoto, Michiyo. "Plight of Fukushima farmers takes its toll." *ft.com*, Japan. At http://www.ft.com/cms/s/0/5df7847e-5a07-11e0-ba8d-00144feab49a.html#axzz3PwApkjh4.

"New prediction shows triple quake to kill 400,000 in Japan." *The Yomiuri Shimbun* (July 11, 2012). At http://www.asianewsnet.net/news-33210.html.

Nietzsche, Friedrich. *The Portable Nietzsche*. Trans. Walter Kaufmann. New York: Viking Press, 1968. At http://www.inp.uw.edu.pl/mdsie/Political_Thought/twilight-of-the-idols-friedrich-neitzsche.pdf.

Norimitsu, Onishi. "Death Reveals Harsh Side of a 'Model' in Japan." *New York Times* (Dec. 19, 2015). At http://www.nytimes.com/2007/10/12/world/asia/12japan.html?hp=&_r=0.

Parker, Bruce. *The Power of the Sea: Tsunamis, Storm Surges, Rogue Waves, and Our Quest to Predict Disasters.* New York: Palgrave Macmillan, 2010, 2012.

Perrin, Noel. *Giving Up the Gun: Japan's Reversion to the Sword, 1543–1879*. Boston: David Godine, 1979.

Perry, Ronald W., and E. L. Quarantelli, eds. *What Is a Disaster? New Answers to Old Questions*. International Research Committee on Disasters: Xlibris Corporation, 2005.

Potter, Polyxeni. "Francisco José de Goya y Lucientes (1746–1828). Self-portrait with Dr. Arrieta (1820)." *Emerging Infectious Diseases* 10 (2004): 974.

Prabhune, Manish. "Rapeseed blossoms to start off the spring season." *Japan Travel.com* (March 2013). At http://en.japantravel.com/view/yellow-flowers-at-hamarikyu-garden.

Prince, Samuel Henry. *Catastrophe and Social Change: Based upon a Sociological Study of the Halifax Disaster*. New York: Columbia University Press, 1920.

Putnam, Robert. *Bowling Alone: The Collapse and Revival of American Community*. New York: Simon and Schuster, 2000.

Rajan, S. Ravi. "Missing Expertise, Categorical Politics, and Chronic Disasters: The Case of Bhopal." 237–59 in *Catastrophe & Culture: The Anthropology of Disaster*. Edited by Susanna M. Hoffman and Anthony Oliver-Smith. Santa Fe, NM: School of American Research Press, 2002.

Riesman, David, Nathan Glazer, and Reuel Denney. *The Lonely Crowd*, rev. ed. New Haven, CT: Yale University Press, 2001; orig. publication 1961.

Rimer, J. Thomas, ed. *Culture and Identity: Japanese Intellectuals during the Interwar Years*. Princeton: Princeton University Press, 1990.

Saint Bonaventure. *De sex alis seraphim* [On the Six Wings of the Seraphim]. Trans. Sabinus Molitor. St. Louis: B. Herder, 1920.

Saint Joseph Edition of *The New American Bible*. New York: Catholic Book Publishing Co., 1970.

Sakurai, Katsunobu. Interview. Minamisoma City, March 19, 2014.

———. "SOS from Mayor of Minami Soma City." At http://www.youtube.com/watch?v=70ZHQ--cK40. Also at https://www.youtube.com/watch?v=R6URqs9kb20.

Samuels, Richard J. *3.11: Disaster and Change in Japan*. Ithaca, NY: Cornell University Press, 2013.

Sansom, George. *A History of Japan: 1615–1867*. Stanford, CA: Stanford University Press, 1963.

Senter for Krisepsykologi. Etterlatte foreldre, partnere, søsken og venner etter Utøya-drapene 22.07.2011: Oppsummering av forskningsprosjekt–2¼ är etter. Bergen (2014). [Center for Crisis Psychology. Bereaved parents, partners, siblings and friends after Utøya murders 22.07.2011: Summary of a research project-2¼ years later].

"*Shinsai kara hantoshi Tōhoku yo—'Tōhoku gaku' o teisō suru minzoku no gakusha. Akasaka Norio-san.*" [Six Months Since Disaster in Tohoku: Folklorist Akasaka Norio Advocates "Tohokuology."] *Asahi Shimbun*. September 10, 2011.

Skilled Veterans Corps of Fukushima. "The fundamental status and capacity of Skilled Veterans' Corps for Fukushima." At http://svcf.jp/english/fundamental_en.

———. "Proposal for the participation of skilled veterans in the cleanup works of the Fukushima Daiichi Nuclear Power Plant accident." At http://svcf.jp/english/proposal_en.

Slater, David H. "Urgent Ethnography." 25–49. In *Japan Copes with Calamity: Ethnographies of the Earthquake, Tsunami and Nuclear Disasters of March 2011*. 2nd ed. Edited by Tom Gill, Brigitte Steger, and David H. Slater. Bern, Switzerland: Peter Lang, 2013, 2014.

Smith, Gar. *Nuclear Roulette: The Truth about the Most Dangerous Energy Source on Earth*. White River Junction, VT: Chelsea Green Publishing, 2012.

Solnit, Rebecca. *A Paradise Built in Hell: The Extraordinary Communities That Arise in Disaster*. New York: Penguin Books, 2009.

Steele, M. William. "The Great Northern Famine of 1905–1906: Two Sides of International Aid." At http://subsite.icu.ac.jp/iacs/PDF/ACS39_PDF/ACS39_01Steele.pdf.

Steger, Brigitte. "Solidarity and Distinction through Practices of Cleanliness in Tsunami Evacuation Shelters in Yamada, Iwate Prefecture." 53-75. In *Japan Copes with Calamity: Ethnographies of the Earthquake, Tsunami and Nuclear Disasters of March 2011*. 2nd ed. Edited by Tom Gill, Brigitte Steger, and David H. Slater. Bern, Switzerland: Peter Lang, 2014.

"Suicide in Japan." *Nippon.com* (Oct. 9, 2014). At http://www.nippon.com/en/features/h00075/

Sulmasy, Daniel P. *The Rebirth of the Clinic: An Introduction to Spirituality in Health Care*. Washington, DC: Georgetown University Press, 2006.

Tabuchi, Hiroko. "An Aging Japanese Town Bets on a Young Mayor for Its Revival." *New York Times* (April 26, 2012). At http://www.nytimes.com/2012/04/27/business/global/aging-japanese-town-bets-on-a-young-mayor.html?pagewanted=all&_r=0.

Talib, Nassim Nicholas. *The Black Swan: The Impact of the Highly Improbable*, 2nd ed. New York: Random House, 2010.

Terkel, Studs. "Hope Dies Last." 105–10. In *The Life of Meaning: Reflections on Faith, Doubt, and Repairing the World*. Edited by Bob Abernethy and William Bole. New York: Seven Stories Press, 2007

Tkach-Kawasaki, Leslie M. "March 11, 2011 Online: Comparing Japanese newspaper websites and international news websites." 109–23. In *Natural Disaster and Nuclear Crisis in Japan: Response and Recovery after Japan's 3/11*. Edited by Jeff Kingston. London: Routledge, 2012.

Tono City Council and Mayor Toshiaki Honda. "Logistical Support in Tono's Coastal Disaster Zone: Community Ties Aimed at Reconstruction through Relationships." March 5, 2014.

Tono Magokoro Net Film Team. "Great East Japan Earthquake Series, Witnesses of the Tsunami, Ryoichi Usuzawa." Tono Magokoro Net (2012). At http://translate.google.com/translate?hl=en&sl=ja&u=http://tonomagokoro.net/&prev=search.

Tsuboi, Yuzuru. "Interview: Anthropologist says Tohoku region must gain independence from Tokyo." *Asahi Shimbun* (August 1, 2014). At http://ajw.asahi.com/article/views/opinion/AJ201408010006.

Tsuchida, Tomoaki. Interview. Tokyo, Japan. March 25, 2014.

Turkle, Sherry, ed. *Evocative Objects*. Cambridge, MA: MIT Press, 2007.

U.S. Geological Survey. "Magnitude/Intensity Comparison." At http://earthquake.usgs.gov/learn/topics/mag_vs_int.php.

U.S. Nuclear Regulatory Commission. "Backgrounder on Chernobyl Nuclear Power Plant Accident." May 2013. At http://www.nrc.gov/reading-rm/doc-collections/fact-sheets/chernobyl-bg.html.

Vicentini, Alessandra. "The Fukushima Nuclear Crisis E-Coverage: A Linguistic Analysis of *Sciencemag.org* and *Scientificamerican.com*." 297-314. In *The Popularization of Specialized Discourse and Knowledge Across Communities and Cultures*. Edited by Susan Kermas and Thomas Christiansen. Bari, Italy: Edipuglia, 2013.

Wakatsuki, Yoko, and Madison Park. "TEPCO ordered to pay compensation in landmark Fukushima suicide case." *CNN* (August 27, 2014). At http://www.cnn.com/2014/08/27/world/asia/japan-fukushima-health-effects/.

Watsuji, Tetsuro. *Fudo: Climate and Culture: A Philosophical Study*. Trans. Geoffrey Bownas. Westport, CT: Greenwood Press, 1988; reprint, orig. publisher Tokyo: Monbusho, 1961.

———. *Rinrigaku*. Trans. by Yamamoto Seisaku and Robert Carter. Albany: State University of New York Press, 1996; orig. published in Japanese by Iwanami Shoten, Tokyo, 1937, 1942, 1949.

Wilhelm, Johannes, and Alyne Delaney. "No Homes, No Boats, No Rafts: Miyagi Coastal People in the Aftermath of Disaster." 99–124. In *Japan Copes with Calamity: Ethnographies of the Earthquake, Tsunami and Nuclear Disasters of March 2011*. 2nd ed. Edited by Tom Gill, Brigitte Steger, and David H. Slater. Bern, Switzerland: Peter Lang, 2014.

Wilson, Tom. "In Japan's Disaster Zone, Recovery Is Just Beginning." *Time* (2011). At http://content.time.com/time/world/article/0,8599,2097511,00.html.

World Nuclear Association. "Chernobyl Accident 1986." Dec. 2014. At http://www.world-nuclear.org/info/Safety-and-Security/Safety-of-Plants/Chernobyl-Accident/.

Yablokov, Alexy V., Vassily B. Nesterenko, and Alexy V. Nesterenko. "Chernobyl: Consequences of the Catastrophe for People and the Environment." *Annals of the New York Academy of Sciences* 1181 (2009).

Yamasaki, Erika. "What We Can Learn from Japan's Early Earthquake Warning System." *Momemtum* 1 (2012): 1-26. At http://repository.upenn.edu/cgi/viewcontent.cgi?article=1022&context=momentum.

Yanagita, Kunio. *The Legends of Tono*. Trans. Ronald A. Morse. Lanham, MD: Lexington Books, Rowman & Littlefield, 2008; orig. *Tono monogatari*, 1910.

Zack, Naomi. *Ethics for Disaster*. Lanham, MD: Rowman & Littlefield Publishers, Inc., 2009.

Index

About the Author

Michael C. Brannigan (PhD, Philosophy, MA, Religious Studies, University of Leuven, Belgium) is the George and Jane Pfaff Endowed Chair in Ethics and Moral Values at The College of Saint Rose in Albany, New York. Holder of the first endowed chair in the college's history, he also teaches at Albany Medical College's Alden March Bioethics Institute and St. Bernard's School of Theology and Ministry in Albany.

Born Kenji Kimura in Fukuoka, Japan, his Japanese mother, Misae Kimura, and Irish-American father, Tom Brannigan, surely prompted his lifelong interest in intercultural and comparative studies. He has an enduring passion for music and philosophy and has engaged meaningfully in work as a hospice volunteer, hospital ethics consultant, and ethics committee advisor.

Michael's specialty lies in ethics, Asian philosophy, medical and intercultural ethics, and phenomenology. Along with numerous articles, his books include: *Ethics Across Cultures*; *Striking a Balance: A Primer in Traditional Asian Values*; *The Pulse of Wisdom: The Philosophies of India, China, and Japan*; *Healthcare Ethics in a Diverse Society* (co-authored); *Cross-Cultural Biotechnology*; *Ethical Issues in Human Cloning*; and more recently *Cultural Fault Lines in Healthcare: Reflections on Cultural Competency*. He serves on the editorial boards of *Health Care Analysis: An International Journal of Health Care Philosophy and Policy* and *Communication and Medicine: An Interdisciplinary Journal of Healthcare, Ethics and Society*.

Aspiring to bridge academia and the general community, he speaks at numerous community, national, and international venues. He also writes a monthly column on ethics for the *Albany Times Union*, New York's Capital District's major newspaper, at http://www.timesunion.com/brannigan/.

Michael and his radiant and beautiful wife Brooke live in Niskayuna, New York. For fun, he reads a lot, plays piano, guitar, tennis, ocean kayaks, and still practices martial arts. His website is www.michaelcbrannigan.com.

Introduction

"Disaster" is one of those terms even experts in disaster research have difficulty defining. What constitutes a "disaster"? For laypersons like myself, the meaning of "disaster" seems evident. We often interchange "disaster" with "catastrophe," "calamity," "tragedy," and other terms. I will occasionally do so in this book. Nonetheless, as to disaster's exact measurements, the jury is still out.

Neil Britton, a disaster professional based in Kobe, Japan, tells the tale of the Chinese emperor who asked his court painter, "What is easy to paint, and what is difficult to paint?" After pondering, the artist responded, "Dogs are difficult, but demons are easy." Why? Since everyone sees the obvious, dogs, everyone believes they know its essence.[1] And we end up with an infinite bundle of beliefs. Hence, it is difficult to capture the obvious. Not so with what is less obvious, indeed rare. Who will challenge what I see and others do not see? Britton argues that this captures the quandary about defining "disaster." Is it that obvious, or, when it hits, is it unique?

For our purposes, I will simplify the question by reducing "disaster" to some bare essentials. These are essentials that most disaster experts, particularly pioneers such as Enrico L. Quarantelli, single out as a disaster's key ingredients, whether that disaster is natural, technological, human-incurred, or combinations thereof. To begin with, a disaster is complex and diachronic: it is both an event (which can be sudden like a tornado, or gradual like famine) and the event's aftermath, how things unfold in terms of response and effect, short and long-term. In other words, a disaster is a drawn-out, protracted process. Much of this has to do with multiple forces that are social, cultural, political, economic, religious, and educational. Each of these carry hidden norms, structures, and values. We can unpack this further by denoting what experts consider the key ingredients of disaster. They include features pertaining to impact, hazards, population, and vulnerability. Since I will more closely address Japan's March 11, 2011, (what many Japanese refer to as "3.11") impact in the next chapter, let us dig further into the conventional components of hazards, population, and vulnerability, and also disaster's trait as a crack in the routine.

ANATOMY OF DISASTER

Hazards

Noted geographer Kenneth Hewitt points out that disasters comprise any one or more of three fundamental types of hazards: natural, technological, and social. Wherever we live, we are generally at risk in dealing with any of these hazards.[2] Natural hazards include risks that are atmospheric, hydrological, geological, and biological. There are areas in the United States that are more susceptible to hurricanes, tornadoes, landslides, earthquakes, firestorms, avalanches, mudslides, or flood zones. If we choose to live sea-level near beaches in Rhode Island (The Ocean State), there is a reason why we pay flood insurance. If we live in upper Oklahoma, tornadoes form part of the natural hazard of living there. If we live in California, depending upon where, we risk the increased possibility of earthquake, mudslide, drought, and wildfires.

Technological hazards include those dangerous, toxic materials we can be exposed to in our environment. If we live near nuclear reactors, chemical plants, power plants, or waste centers, we face a variety of health-related risks.

Social hazards are risks we face that are human-incurred, as in areas of civil conflict, terrorism, war, and where there is deliberate use of hazardous materials. Living in areas that are heavily populated, especially if we live in ghettos, or in areas of poverty, or in places ruled by corrupt government and exploitative industry exposes us to these social hazards.

Again, disasters involve hazards that are natural, technological, and social. Philosopher Naomi Zack incorporates these features, adding elements of large-scale casualties and unexpectedness, in her examples of disasters:

> A *disaster* is an event (or series of events) that harms or kills a significant number of people . . . Disasters include . . . fires; floods; storms; earthquakes; chemical spills; leaks of, or infiltration by, toxic substances; terrorist attacks . . . ; epidemics; pandemics; mass failures in electronic communications . . . Disaster always occasion surprise and shock . . . .[3]